ENVIRONMENTAL HARASSMENT
OR
TECHNOLOGY ASSESSMENT?

by

DEREK MEDFO

Department of Econometrics and Math

INTERNATIONAL INSTITUTE
FOR THE MANAGEMENT
OF TECHNOLOGY

INSTITUT INTERNATIONAL
DE GESTION DE LA
TECHNOLOGIE

INTERNATIONALES INSTITUT
FÜR FÜHRUNGSAUFGABEN
IN DER TECHNIK
MILAN, ITALY

ELSEVIER SCIENTIFIC PUBLISHING COMPANY

AMSTERDAM / LONDON / NEW YORK

1973

ELSEVIER SCIENTIFIC PUBLISHING COMPANY
335 Jan van Galenstraat
P.O. Box 1270, Amsterdam, The Netherlands

AMERICAN ELSEVIER PUBLISHING COMPANY, INC.
52 Vanderbilt Avenue
New York, New York 10017

Library of Congress Card Number: 72-97431

ISBN 0-444-41110-0

With 21 illustrations and 6 tables.

Printed in The Netherlands

Preface

The environment as a whole is divided into three parts: ecological, social, and economic.

An author would be hard pressed to find a riskier subject than one with an apparent misnomer for a title and a rapidly growing, multi-disciplinary, band of adherents. I have willingly undertaken the hazardous task of writing on technology assessment because this amorphous subject has already started to spread its tentacles around business, social, and governmental activities. Indeed, it has been claimed by Carpenter (1972) that

> "Technology assessment is a social invention which will facilitate appro-
> priate societal control of physical inventions. Technology assessment is a
> tool for the renewal of our basic decision making institutions, the demo-
> cratic process and the free market economy. It is, not surprisingly, being
> developed just in time — remember that necessity is the mother of inven-
> tion."

To make the last quotation intelligible, this book should begin with definitions but, unfortunately, there are difficulties. In the absence of a unanimously agreed definition of technology assess-ment and the glossary of new "words" that have arisen during its practice, it is not always possible to avoid some uncertainty of meaning. Even experienced journalists have been perplexed by the ambiguity of language used and one has written the following, in Innovation (1971)

v

"Technology has been variously used as a synonym for the sciences (both 'hard' and 'social'), research, invention, product development, et al. Technology assessment has been taken to mean the process of specifying and evaluating the first-, second-, third-, and nth-order side effects of all the above on the environment.

The environment, it should be added, is another catch-all term used as a synonym for virtually every physical and social factor influencing human experience."

Nevertheless, comprehension of the drift of this book may be obtained (even with only cursory reading) if the reader accepts the following, thoughtful, definition made by Huddle (1972).

"Technology assessment is the purposeful, timely and iterative search for unanticipated secondary consequences of an innovation derived from applied science or empirical development, identifying affected parties, evaluating the social, environmental, and cultural impacts, considering feasible technological alternatives, and revealing constructive opportunities, with the intent of managing more effectively to achieve societal goals."

World-wide support for, and interest in, technology assessment has inevitably arisen at a critical time when the human race is neither master nor slave of technology and we are apparently faced with increasing pressures of population growth, voracious demands for, and the possibility of irreversible despoliation of, terrestrial reserves of natural resources, instabilities in social, ecological, and economic environments, and increasing dependence on technology.

Consequently, in countries with well-developed industry and time to devote to other than immediate problems, much controversy has arisen about the assessment of social, ecological, and economic impacts of technology. Constructive disputes about assessment methodology are common and would be of first importance, except that it is significant to note that arguments about the structure of organizations to be trusted with assessments are being pursued with at least equal vigour. Simply expressed, the most clearly demarcated arguments run somewhat as follows.

(1) Technology will find its own acceptable level in the market and any disbenefits, arising from the use of a technological product, will react adversely on its success.

An extension of this argument postulates an ameliorating feedback effect within society which obviates the need for a technology assessment function outside the industrial system.

(2) Present procedures of marketing virtually result in an eternal struggle for advantage, with consumers at a continuing disadvantage because of their solitary (essentially non-technical) roles. Con-

sumer councils and advocates are then proposed as a means of enhancing the feedback, postulated in the first argument, through the mechanism of banding consumers together to make collective assessments of technological goods (and services) and their side effects.

(3) Side effects are only perceived *after* consumption has largely taken place and damage been done. This argument is often strengthened by suggesting that many additional benefits of technology have been foregone, by rushing into improvised means of satisfying demand, through the application of inferior technological solutions. In either case, it is postulated that technology assessment should contain an element of prediction of future consequences. Furthermore, so that demarcation of consequences may be impartially made, it is also argued that the assessment function should be carried out by organizations which are independent of industrial, commercial, and other parochial interests.

This book discusses matters cogent to the above arguments. Specifically it contains

(1) an examination of the recent growth of technology assessment activity,

(2) a description, and critique, of some of the known methods of technology assessment, and

(3) a discussion of social, economic, and legal consequences of legislating for an overt assessment procedure. The text is not exclusively concerned with quantitative techniques although these are necessarily discussed in Chapters 5—8 inclusive. Non-mathematical readers may skip the numerate details and read only the verbal critiques of the quantitative methods described. But it is not recommended that this decision is lightly made without first ascertaining that the mathematical detail is indeed beyond comprehension because technology assessment is usually distinguished by the fact that most of the mathematics used is of no higher standard than that taught to school leavers. Invariably, technology assessment has a pseudo-scientific appearance despite the comprehensibleness of its inherent logic and the last advice is not entirely gratuitous. It should be realised that a tendency towards pseudo-scientific presentation can stem from a desire to achieve plausibility or, worse, obscurity. Even without the prompting of uncomfortable thoughts on ecological or socio-economic disaster, all readers should try to familiarize themselves with the analytic aids available to their elected (and non-elected) decision makers. This last exhortation is of some importance since ultimate arbiters of

decision rely upon assessments when they are about to promote or condone environmental change.

A major purpose of this publication is to generate more European public awareness of the way in which one of the most crucial elements interacting with our society is, or should be, evaluated. All who would aspire to change or maintain facets of modern societal structure ought to have some understanding of technology assessment. An anciliary purpose is to generate European interest in the setting-up of Offices of Technology Assessment for national communities or the European Community, preferably the latter.

The U.S. House of Representatives has already passed Bill HR 10243[1] to establish an Office of Technology Assessment (OTA) for the U.S. Congress

> "as an aid in the identification and consideration of existing and probable impacts of technological application"

and the European Economic Community (EEC) should start to consider the relevance of this legislation to European activities. Throughout this book I shall discursively argue that there is a need to accompany overall EEC actions on scientific and technological matters with efforts made by a European OTA, which must have a fair measure of independence from the executive arm of government and serve only (in conformity with the American precept) to extend the information-gathering function of the European Parliament.

Representative Symington of Missouri, when supporting Bill HR 10243, claimed that it was not exactly an "America-first type of Bill because ...... it is clear that West Germany, Britain, Japan, Sweden, Canada, and, in fact, most of the major developed countries of the world have established such an office .....". Nevertheless, most (if not all) of the organizations Symington cited are directly responsible to ministries of trade or economy and only serve parliaments indirectly through the executive agencies of government. Thus there is an unsatisfied European need for advisory OTAs to give advice on complex technological matters directly to parliaments so that politicians and the public are not absolutely dependent upon expertise provided by executive agencies of government and other organizations with vested interests.

EEC collaborative projects will ultimately depend upon more

---

[1] HR 10243, 92nd Congress, 2nd Session, Feb. 9th 1972. See Congressional Record, House, H865 to H886.

mature judgement and become less dependent upon precursor American evidence of technological problems. Too much reliance on advance warning from the U.S.A. is manifestly bad and, as the European Community becomes technologically more experienced, such a dependence will diminish to be replaced with other, more balanced, kinds of technological collaboration. However, the significance of the current American legislation on technology assessment should not be ignored by the "new" Europeans and to facilitate the learning process, the possible lessons we can learn from the recent American experience are discussed from a European viewpoint in Chapter 2.

The subconscious impetus behind my writing is for the reader alone to judge. My conscious reasons are as stated and my only qualifications for presuming to write such a book are that after many years employment as engineer, mathematician, and physicist I have, for the past half decade, worked on technology assessment for the U.K. Department of Trade and Industry and the United Kingdom Atomic Energy Authority. Through the altruistic support of these employers and the generosity of Nuffield College, Oxford, I have had the privilege of mulling over the topic and keeping up with developments in American legislation without the exigencies of day-to-day business.

Because technology assessment is such a new and rapidly expanding subject, it has seemed useful to employ extensive reportage when describing current trends. I hope this semi-journalistic approach has not left me unjustly vulnerable to the charge first made by Robert Burton (1577—1640) in his work Anatomy of Melancholy,

"They lard their lean books with the fat or others' works".

ACKNOWLEDGEMENTS

Without the advice of past and present colleagues, this subjectively biased book could never have been written. I am extremely grateful to the Warden, Fellows, and Students of Nuffield College for the tolerance they showed as I attempted to metamorphose from a physical into a social scientist, the late Sir Ioan Gwilym Gibbon for his support of the Gwilym Gibbon Research Fellowship, and Richard Carpenter and Franklin P. Huddle for the hospitality and guidance they gave me when I worked for a short time in the Congressional Research Service, Library of Congress.

References

Carpenter, R.A. (1972). "The Scope and Limits of Technology Assessment", *OECD Seminar on Technology Assessment, January 26—28, Paris.*

Huddle, F.P. (1972). *A Short Glossary of Science Policy Terms*, Science Policy Research Division, The Library of Congress, preliminary draft, not yet published.

Innovation (1971). "Making Technology Assessable," *Innovation*, No. 25, October.

Contents

Preface   V

Chapter 1. The recent growth of technology assessment . .   1
           Background growth   1
           Growth of European interest in technology
              assessment   2
           The growth of technology assessment to combat
              vested interest   8
           "Invisible colleges" and growth of technology
              assessment 10
           Growth of technology assessment through
              changes in advocacy procedure 13
           References 14

Chapter 2. Technology assessment in the U.S.A. 15
           Introduction 15
           The National Environmental Policy Act and
              technology assessment 16
           The American concept of an Office of
              Technology Assessment 32
           The law and technology assessment in the
              U.S.A. 44

American Industry and the Office of Technology
Assessment 51
References 55
Appendix 1. S.2302 58
Appendix 2. H.R.10243 77

Chapter 3. Technology assessment in the U.K. public
service 92
Introduction 92
Does the service ever explicitly consider
technology impact? 93
Comments 95
Lord Rothschild's emporium of science 97
Conclusions 101
References 102

Chapter 4. The British Programmes Analysis Unit (PAU) . 103
Comments on the PAU 103
The external contacts made by the unit 114
Attitudes of the unit's staff to their past
experience 116
A distillation of lessons learnt by the unit . . . 118
References 121
Appendix. Published reports and papers by PAU
authors 121

Chapter 5. Benefit—cost analysis in technology assessment . 124
Introduction 124
Analyst—arbiter relationships in benefit- cost
analysis 125
Present value 127
Criteria 131
Probability of achieving present value 134
Sensitivity analysis 145
Comments on mathematical modelling in
benefit—cost analysis 161
Appendix 1 163
Present values of cash flows 163
Numerical procedure for calculating net present
value of non-analytic cash flow curves 170
An example of the use of the numerical method . 171
Appendix 2 174

A detailed example of the use of sampling
  function in benefit- cost analysis 174
References 177

Chapter 6. Cost—effectiveness analysis and technology
  assessment , ... 178
Introduction 178
An example of the cost—effectiveness of
  computers 180
Concluding comments 195
Reference 195

Chapter 7. Technological forecasting and its place in
  technology assessment 197
Introduction 198
Exploratory techniques of technological
  forecasting 199
Normative techniques of technological
  forecasting 210
A European relevance technique 211
Annex A 221
Combining rankings with respect to different
  attributes 221
Annex B 225
Cost—risk factor and economic factor in the CPE
  work 225
An American relevance technique 227
Conclusions about the French and American
  relevance techniques 232
Other methods of technological forecasting . . 237
Summary 243
References 245

Chapter 8. Technology assessment of curiosity-motivated
  research 247
Introduction 247
Difficulties of evaluating COR 248
Retrospective study 250
Notional apportionment of benefit 256
Prospective study 265
Foreseeable breakthrough for the research
  sector 267

Feasibility of an X-ray holographic microscope . 275
Possible benefits of X-ray holography 279
Possible links between current research and a
normative COR goal 282
Concluding comments on the prospective
evaluation of COR 283
References 286

Chapter 9. Ethical industrial intelligence 288
Introduction 288
Ethical intelligence 289
Intelligent collection and collation of
information 294
General advice on the organization of
intelligence in technology assessment 300
Concluding comment 303
References 303

Chapter 10. Universities and technology assessment 304
Introduction 304
Systems universities 305
University teaching of technology assessment . . 310
Future action 314
References 317

Chapter 11. Future public involvement in technology
assessment 319
Introduction 319
Portents of future development in the style of
public discourse 320
Categorization of types of environmentalists . . 322
Suggestions for a phased programme of
environmentalist action 325
Conclusion 327
References 327

Chapter 12. Future involvement of industry in technology
assessment 329
Introduction 329
Assessment contracts for industry 330
References 337

Chapter 13. Loose ends 337
     Introduction 337
     Principles 337
     References 351

Index 353

CHAPTER 1

The Recent Growth of
Technology Assessment

*"We are much beholden to Machiavel and others, that
write what men do, and not what they ought to do."*
Francis Bacon, *Advancement of Learning*

Background Growth

Assessment of technology is not a new activity. Since efforts
were first made to harness the laws of nature, man has naturally
been concerned with the consequences of applying new technology. Regrettably, this concern has frequently been confined to the
prediction of benefits likely to accrue to a small sub-set of society.

Innovators have often sought mastery over others through the
calculated use of technology in situations of conflict, as witnessed
by the abnormal growth of new technologies during wars and
periods of intense industrial competition. In truth, during this
century the phenomenal growth of interest in assessment technique owes as much to military "defence" organizations as it does
to the increasing complexity of technology. Evidence can even be
collected which would suggest that the boom in high-technology
correlates well with increased activity of the defence organizations.

There are interesting developments in the mathematics of risk

1

analysis which pay attention to the "trade-off" between increased expenditure of resources of analysis (used for improving the calculation of probability of risk) and the consequences of not taking action before acceptably accurate estimates of risk are made. However, common sense suggests that the greater the threatened risk, the greater will be the amount of analysis devoted to defining the risk accurately and the means of averting the threat. It is a truism to say that increments in the degree of perceived risk, threatened by external forces, give rise to enhanced activity by the assessors and innovators of technology. An equally obvious corollary is that if a sector of society is convinced by a persuasive seer of some real (or imagined) threat there will be a spurt of analytic activity. Recent examples are

(1) the "watershed" during the Second World War when Operational Research (which is a prime source of much technology assessment technique) became a necessary and acceptable discipline,

(2) the rapid growth of Management Science immediately after the War which was engendered by the need for industrial reconstruction and the advent of the digital computer, and

(3) the current bustle of activity in evaluation of the impact of technology which has been prompted by predictions of environmental and ecological disaster.

Notwithstanding the end of the Second World War, the gargantuan weapons of destruction lamentably led to the perpetuation of real threat and an escalation of armament technology and its concomitant assessment. This last perturbation gave rise to continuing military cross-fertilization of civil assessment of technology, nurtured by the military development of more powerful digital computers which increased the calculating power at the civil analyst's elbow (if not his discrimination).

Growth of European Interest in Technology Assessment

For the past few decades, reputable organizations in non-European countries, such as RAND, Stanford Research, Batelle Memorial, and Hudson Institutes have been concerned with "think-tank" activities and the innovation of analytic aids for assessors of technology and makers of scientific policy. These efforts of foreign planners had for some time been kept under surveillance by European observers (particularly at places like OECD) but, until 1967, the only U.K. reaction was phlegmatic and attempts were not made to introduce such techniques into the British (or

European) system. Flair and amateurism sufficed and were conveniently buttressed by operations research which was mainly used to tackle current problems.

Recently, several cumulative effects have created a more favourable European environment for the introduction of new analytic aids to decision making. For example, amongst these effects in the United Kingdom are

the varying state of Britain's economy,

the Fulton Committee's recommendations for a more professional Civil Service,

the "mushrooming" of business school education,

the growth of long-range cum futurist societies, and

the efforts of the communication media, a good example being Kit Pedlar's BBC (TV) series "Doomwatch" which portrays a bureau of scientists on the qui vive for disastrous side-effects of science and technology.

The general growth of interest in the new pseudo-science of technological forecasting (or futurology) can be dated, in Europe, to the appearance of books by Kahn and Wiener (1967) and Jantsch (1967), although precursors of this general interest abound in European literature (H.G. Wells, Jules Verne) and the work of dedicated professional groups like de Jouvenal's team working on "Futuribles" (see de Jouvenal, 1963).

Probably the most commonly accepted definition of this predictive element of technology assessment has been given by Erich Jantsch.

"Technological forecasting is the probabilistic assessment, on a relatively high confidence level, of future technological transfer."

Two generally accepted forms of technological forecasting have become recognized: exploratory forecasting, which is a form of temporal extrapolation from a base of retrospective fact, and normative forecasting, which attempts to assess future goals, needs, desires, missions, etc., and works backwards to the present identifying what changes are required on the way. Inevitably there are considerable dangers in both and the use and abuse of technological forecasting will be dealt with in Chapter 7. But to proceed with this discourse, it is necessary to say something of these dangers in this chapter. In exploratory forecasting there is a danger that future discontinuities, which act as branch points in the growth of science and technology, are assumed to be smoothed out, and policy based on such a forecast could stifle progressive ideas. In

3

normative forecasting, potentially greater danger lies in the identification of mission-orientated goals desired by some sectional interest and the assumption that these goals are compatible with collective social goals.

Technological forecasting has been practiced in Europe for not less than five years on a fairly wide scale in industry and the public sector and this experience will, as already said, be examined later. However, apropos of the European growth of technology assessment activity, it is requisite to point to a "threat" phenomenon which exists within the process itself. Early in 1968 two European meetings heralded the repetitive staging of conferences on the subject of technological forecasting, the 1st European Conference on Technological Forecasting (see Arnfield, 1969) and the U.K. Conference on Technological Forecasting (see Wills, 1969). Despite the naivety of the techniques presented at these conferences and the natural sceptism of the conference delegates, an apparent threat to establish business and bureaucratic "pecking order" was generated. Influential men left the conferences with a feeling that there might well be some advantage to be gained from the practice of the new art. From this thought grew the suspicion that competitors might, pre-emptively, grasp the potential advantages unless the man's company, or department, started to make (or buy) its own technological forecasts. This rationale helped prompt the European boom in technological forecasting and it was possible to witness an interesting phenomenon where *the inception of a "new" method created an enhanced probability of competitive-threat and thus led to a general attempt to use the new method.* (A cynic might, at this point, allege that some unscrupulous purveyors of management science know this phenomenon well and can calculate, to a nicety, the time it takes for general rejection of a new "ism".) Restraint, however, ultimately prevailed and despite the subject having its own built-in, reinforcing, positive feedback, the unimaginative use of the more obvious packaged techniques yielded sufficient warning that technological forecasting might be used with considerable discrimination and finesse.

It is now generally realized that technological forecasting is no substitute for, but is a useful adjunct to, effective intelligence activity in technology assessment. The growth of technological forecasting activity is now proceeding asymptotically to an acceptable plateau. Moreover, a basic tenet of extrapolatory forecasting is that when progress has been made along an S-shaped curve to some upper bound of growth, the inception of another S-shaped

4

curve (which starts around the asymptote of the old S-curve) is imminent (see Chapter 7). The majority of informed opinion now predicts a new growth in the practice of "ethical" industrial intelligence as opposed to unethical industrial espionage which has, of course, been with us for many years (see Chapter 9). Ethical industrial intelligence allied to technological forecasting, and other assessment techniques is expected to show considerable benefit to practitioners. Digital computers are also expected to boost this old intelligence activity, which has become newly respectable, and overtly popular, by providing efficient data-banks (see Chaumier, 1972). One would also expect that, as industrial intelligence activity increases the total information available to the assessor of technology, the use of mathematical modelling will also increase since modelling is one way of systematically handling the anticipated flood of multi-variabled data.

Another important factor, at least in the U.K., which has contributed to the growth of technology assessment is the involvement of government in industrial affairs. (Even governments who would disengage themselves from too direct an involvement with industry enhance the growth of technology assessment because they like to present persuasive analyses explaining why they take certain actions!)

At a press conference, held on 15th January 1969, the British Minister of Technology made the following statement.

"A basic objective of the Ministry is to strengthen the competitive ability of British industry by accelerating the rate of application of innovation."

The Minister then went on to announce the following actions.

(1) The setting up of ten Research Advisory Committees where industrialists will have a say about the links between government research resources and industry.

(2) Steps to foster direct links between the Ministry of Technology establishments, universities, and industrial organizations.

(3) Authorisation of the Atomic Energy Authority to undertake research and development (R & D) outside the nuclear field.

(4) Arrangements for the work of Mintech defence establishments to be made available for civil exploitation, without condoning the use of staff no longer required on defence or civil aeronautical work.

(5) The acceptance by Mintech establishments of more contract work from industry.

(6) Arrangements for closer association of the National Re-

search and Development Corporation with the civil exploitation of defence research.

(7) Research programmes funded by Mintech were to be subjected to rigorous analysis by the newly formed Programmes Analysis Unit (PAU) which would *"assess the prospects of profitable exploitation, use economic and market surveys as well as forecasts of future environment"*, and work closely with industry and universities to get the best advice available.

More recently, in November 1971, the Lord Privy Seal presented to Parliament Cmnd. 4814 ("A Framework for Government Research and Development") in the form of a Green Paper containing two contradictory reports by Lord Rothschild, Head of the Cabinet Office's Central Policy Review Staff (CPRS) and by a Sub-committee of the Council of Scientific Policy chaired by Sir Frederick Dainton. This Green Paper was intended to trigger discussion and debate amongst the scientific community, and others who have an interest in the application of scientific and technological innovation, before the Government reached a decision on the detailed application of Lord Rothschild's "customer—contractor" role-playing system for the control of government sponsored R & D. Lord Rothschild's recommendation is that Chief Scientists be deputed to represent "customers" and Controllers of R & D be deputed to represent research services as controllers and vendors of R & D. Without wishing to enter here the controversy about the efficacy of such role-playing (see Chapter 3), it is quite clear that if Rothschild's system is to be properly implemented, there will be yet a further increment of activity in technology assessment simply because true dichotomy of responsibility between Chief Scientists and Controllers of R & D will lead to a partial duplication of a very time-consuming function, that of evaluation and assessment!

In order to advise in depth, rather than at the specious level of what research has currently revealed, on potential means of satisfying needs with techniques and products from the forefront of applied research, the Chief Scientist's staff will need to be continually on the look-out for means of satisfying needs whereas the Controller of R & D and his staff, as purveyors of R & D services, will be continually searching for needs which they can satisfy with the means at their disposal.

It is not the purpose of this book to examine in useful detail the efficacy of the Ministry of Technology and the Department of Trade and Industry, or the practicality of Lord Rothschild's pro-

6

posals. These subjects have been touched on in order to give some idea of the developing rationale which lay behind the creation of the British PAU (see Chapter 4) and to indicate the possibility that such technology assessment units could become manifold. At the onset of this governmental interest in technology assessment, the announcement of the PAU's central evaluation function had a significant effect on some British industrialists and postulants for R & D funds for two major reasons.

(1) First, whatever convictions were held in the late 60's about the Government's "direct interference" with industry, organizations trying to evaluate the likely outcome of increased collaboration with Government saw some merit in trying to understand the methods of assessment, which were being encouraged in the PAU.

(2) Secondly, there was a possibility that the analysts, working in the public service, would be better able to develop and scrutinize techniques of assessment because they had a central position and access to wider information. (Some unkindly said the civil servants also had time to waste chasing the more esoteric realms of assessment technique.) Those who held to the last reason were reinforced in their belief by an announcement that the relevant PAU functions were

(1) to develop technical and economic criteria and techniques of analysis by which the potential benefits of R & D programmes can be assessed,

(2) to apply these criteria and techniques to assessment of current and proposed R & D programmes, and

(3) to provide a focal point of expert knowledge about R & D evaluation on which the Ministry of Technology and the United Kingdom Atomic Energy Authority can draw for their own purposes.

Chapter 4 contains a discussion on the experience of the PAU when it tried to provide a large Ministry of Technology and other Government departments with a technology assessment service. In Chapter 4 a published list of external liaisons made by the PAU with organizations in the public and private sectors is given which is indicative of the contribution made by government to the growth of technology assessment in industry and elsewhere. Subjective comments on the proposals made by Lord Rothchild in Green Paper Cmnd. 4814 are also made in Chapter 3.

The Growth of Technology Assesment to Combat Vested Interest

To explain the most recent (and possibly strongest) fillip to growth of technology assessment, it is necessary to say something about the abuse of analytic techniques when they are used to brief arbiters of decision. Abuse is possible because of the established respectability of the "scientific method". Ascientific arbiters of policy have long been accustomed to accepting facts revealed by the analysis of their scientific colleagues and unfortunately the "sciento-mathematic" language of technology assessment (particularly its sub-disciplines technological forecasting and mathematical modelling) is almost indistinguishable, to the generalist, from that of rigorous science. Sometimes there is no need for the arbiter to make this distinction because isolated examples of "scientific method" do sometimes occur in assessments. But, unfortunately, there are more cases where discrimination is difficult because real and facsimile applications of the scientific method lie side-by-side in the same assessment. Arbiters cannot be heavily criticized for putting faith in technology assessments which are apparently begotten by the scientific method. In practice, arbiters very often accept assessments as accurate and confine their judgement to the desirability of the political, social, and economic outcomes predicted by the assessors without making sufficient challenge of the assessors methods of analysis. Arbiters very often feel that they have done their duty after struggling with the assessor's predication of outcome and selecting, from the apparently cautious presentation of the probability of occurrences, a likelihood of one of the outcomes occurring. Nevertheless, arbiters usually find it extremely difficult, or fail, to discover the causality assumed by the analyst and regrettably, at times, to detect the analyst's value judgements that have not been explicitly revealed and lie buried in the analysis. (Any reader who doubts the truth of the last statement should play the arbiter's role when reading *Limits to Growth* by Meadows *et al.*; see the next section on "invisible colleges".) Before the advent of so many diverse analytical aids, persuasiveness was quite often legitimately built into the final presentation of assessments, if only to convince the recipient that a good assessment was acceptable. Now, unfortunately, the rate of inception of new assessment techniques is so fast that the process of widespread critical appraisal of new methodologies has failed to keep up and, as with all new disciplines, a disproportionally large number of unproven techniques is currently in use. Thus it is possible, if there is a will to add plausibility to requests for funds, or other support, to base an assessment on a bogus forecast or a piece of analysis obtained with a "flexible" technique. Proffered arguments can be consciously biased but the analyst's integrity is such that bias is more likely to be subconsciously generated by the pressures upon him.

Conversely, a protagonist may well level the last accusation against an intellectually honest attempt at assessment made by a protagonist of a rival viewpoint who has been forced to advise on an imminent decision in the absence of adequate information.

All such conflict tends to make technology assessors act, more and more, as advocates for particular causes and leads to further proliferation of assessment as a defensive and offensive technique. Even if all the assessment techniques now in use were acceptable to all, advocacy of technological causes would still occur because there will always be genuine areas of uncertainty which call for subjective prediction. Technology assessment can never be more than an art with strong support of scientific method and room for irreducible disagreement. Optimistically, one might hope that the residue of disagreement will be made considerably smaller through more rigorous questioning of the assessors output by elected arbiters of decision and by the public. If assessors increase their advocacy role under such surveillance, they might even be expected to refine some of the loose techniques currently in vogue.

In the present state of the art, even good analysts find considerable difficulty in understanding, and appraising, the work of their colleagues, especially when effort has been made to collect and selectively present evidence slanted towards a particular objective. Therefore, to protect collective interest, the need has been perceived for a central agency owing no direct allegiance to particular sub-sectors of the community (including executive agencies of government) which can operate without bureaucratic impedance in the field of technology assessment[1]. *Ralph Nader type "whistle-blowing" (the encouragement of anonymous revelation, supposedly for the public's benefit, of confidential information) is not the only way to safeguard against the bureaucratic impedance which is sometimes inserted by administrations for parochial rea-*

---

[1] Schlesinger (1967) has noted that government agencies tend to withold information that puts them in an unfavourable light. They often induce a certain fuzziness, if not distortion, in order to maintain their competitive positions in relation to other agencies.

sons. Central assessment agencies may be a more constructive way of canalising activists who are socially aware of the harmful side-effects of some technologies.

Central Offices of Technology Assessment, with responsibility only to legislative branches of government and the public, if they are established nationally (as for the proposed American OTA, see Chapter 2) or for specific economic blocs[1], will represent a plateau in the growth of technology assessment as it is presently understood. Far-sighted optimists even see future political bodies comprised of people experienced in the analysis of first- and second-order impacts of technology on the economic and social structure of society. More pragmatic realists will surely expect, as a minimum, that politicians and administrators will not confuse second-order beneficial effects with first-order disbenefits.

"Invisible Colleges" and Growth of Technology Assessment

American attempts to set up an Office of Technology Assessment have only begun to reach fruition after nearly a decade of effort. Politicians cannot, therefore, be expected hurriedly to enact appropriate statutes to sponsor OTAs. Such inevitable procrastination leaves unsatisfied a need which privately sponsored, non-profit making, organizations readily compete to fill.

One notable "do-gooder" amongst the private organizations willing to don the mantle of guardian-of-technology voluntarily is The Club of Rome. This club was formed by a multi-disciplinary group of thirty individuals who met in the Academia dei Lincei, Rome, in April 1968. Within the definition given in the preface to this book its members are actively concerned with technology assessment because The Club of Rome's stated purpose is to

> "foster understanding of the varied but interdependent components — economic, political, natural, and social — that make up the global system in which we all live; to bring understanding to the attention of the policy makers and the public worldwide, and in this way to promote new policy initiatives and actions".

---

[1] "Each government should establish, a special structure, at Ministerial level or in a manner independent of the Executive, a special structure that would be responsible for anticipating the likely effects, threatening or beneficial, of technological initiatives and developments." OECD (1971), *Science Growth and Society.*

Another privately inspired organization, bent on satisfying the same public need, is the International Society for Technology Assessment (ISTA), which announced in March 1972 that

"a group of foremost international scientists has joined with Alvin Toffler, author of the best seller *Future Shock*, to form an international society to assess the consequences of technology".

Both of these "invisible colleges" will tackle the almost insurmountable problem of finding democratic means to preserve and improve the quality of human life. Presumably, they will work through conferences, seminars, commissioned research, and the publication of relevant books and tracts. The Club of Rome has about seventy members and is consequently somewhat elite. ISTA has some equally distinguished members but intends to generate a membership of thousands and is, despite the Society's rather loose use of the word "scientist", to comprise nearly all academic disciplines, housewives, schoolchildren, and other interested groups.

Future influence of these "invisible colleges" on technology assessment is not easy to predict. ISTA is due to hold its first international conference in The Hague, The Netherlands, sometime in 1973 or 1974. Comments and judgement have to be deferred until then but, meanwhile, Chorus (1972) is available to interested parties. The Club of Rome has been somewhat quicker off the rostrum and has already been extensively featured by the news media because of the catalytic (or scaremongering) book *Limits to Growth*, see Meadows *et al.* (1972), published under their auspices.

Reviews of *Limits to Growth*, which forecasts various dangerous singularities in population, pollution, etc., have not been entirely favourable. Some critics believe that Meadows' communication about envisaged problems facing us, the Planet's "problematique", are constructive and welcome. Other critics, unfortunately, claim that the results of Meadows' computer-programmed models of the world system, made after the style of Professor Jay Forrester, Massachusetts Institute of Technology (see Forrester, 1958, 1968) savour of "doomsday" prediction since detailed exposition of the underlying causality of the model is missing from the Club's first encyclical book. Perhaps after Meadows and his co-authors have digested and replied to the criticism received we shall be better able to judge the usefulness of their book.

Presently, all that can be said about the "invisible colleges" and their contribution to the growth of technology assessment is that they are certainly popularizing the subject. Some are bringing assessment into public favour and others are providing ammunition which may be used to vilify altruistic assessment activity. Personally, I am convinced that both the Club of Rome and the International Society for Technology Assessment have attracted sufficiently meritorious memberships to ensure that their contribution to public debate will ultimately be beneficial. But I do hold the reservation that private organizations, backed by limited donations, cannot fill the major need for nationally sponsored OTAs answering directly to parliaments.

Whether or not OTAs are sponsored, there is obviously a useful role the private organizations can play. They may represent one means of pre-digesting assessments released (by executive agencies of government, industries, and OTAs) to the public. Who knows? perhaps the "invisible colleges" of technology assessment will one day proliferate, conduct their analyses according to political principles, and become the political parties of the future. Non-governmental organizations which are assessment conscious will certainly do much good by smearing existing political boundaries, for example

"sometime in the course of 1972 an International Institute of Applied Systems Analysis will be born on one or two splendid sites already offered, in Vienna or Paris........ Some nine or ten nations will be its founders. Scientists from the Atlantic to the Urals and from both sides of the Pacific will come together .... to determine the state of the art of computerised systems analysis applied to environmental problems the world over, prepare a glossary of neologisms (new environmental words, translated into the necessary working languages) and to analyse "real life" dilemmas such as urban sprawl and industrial pollution".

Claire Sterling (1972)

It is useful to remember that Robert Boyle, Isaac Newton, and others used to hold weekly discussions in what was called, in the early XVIIth Century, the "Invisible College". Indeed, Mueller (1970) reminds us that this organization initiated enquiries into social statistics, the behaviour of man, and population theory. The Club of Rome and its confreres have before them this three-hundred year old precursor as an example. If they can raise their standards of enquiry to those maintained by the Royal Society, which is what the precursor has become, without becoming institutionalized, much can be achieved.

Growth of Technology Assessment through Changes in Advocacy Procedure

A possible revolutionary development in assessment procedure, mooted by Kantrowitz (1972) when giving evidence to a Congressional Committee, could stem from the establishment of OTAs. Dr. Arthur Kantrowitz, Director of Avco Everett Research Laboratory, considers that the key problem which has to be faced by any Office of Technology Assessment is that its need is based on the existence of scientific controversy which surrounds many of the issues with which politicians must deal. He suggests that an OTA must provide a mechanism for producing a statement of the facts of high presumptive validity when there is a scientific controversy about the facts. The techniques suggested by Kantrowitz for increasing the presumptive validity of statement of fact (in the presence of genuine disagreement) involve

(1) identification of scientific advocates who represent the best available talent on both sides of the question and have no duty to be unprejudiced,

(2) these scientific advocates, together with the OTA, would choose a panel of distinguished scientists (not active in the field under discussion) who are thoroughly conversant with the rules of scientific evidence to act as judges,

(3) this panel would then hear the cases presented by the advocates, and the cross examination of those cases by the advocates, and arrive at a statement of the scientific facts which would have a high presumptive validity.

Kantrowitz rightly claims that such a procedure would differ very sharply from the scientific advisory procedures which are in wide use to-day and which would no doubt still continue alongside an OTA. Although it could be argued that if future OTA staff have adequate scientific prestige and the non-vested interest of Kantrowitz's "scientific judges", his Anglo-Saxon judicial procedures would not need to be formalized. In his own words, Kantrowitz's proposition assumes

"a clean separation between the scientific and non-scientific components of a question"

and in technology assessment this is not usually possible. However, notwithstanding this last difficulty, his proposal is no more preposterous than the current system where men of legal background judge between adversary claims involving science and technology

13

when the bench has only vestigial support from expert witnesses. The expectation is that an embryonic OTA would content itself with the collection and analysis of objective facts, and the publication of its findings, so that Parliaments and, whenever possible, the public could show preference according to the established rules of democracy.

Kantrowitz's proposal may one day be implemented in, I suspect, a way that involves "scientific judges" and judges from other disciplines. But not for some time and certainly not before society has been cajoled by "invisible colleges", has accepted the envisaged OTAs, and come to accept the real need for democratic assessment and applications of technology.

References

Arnfield, R.V. (1969). *Technological Forecasting*, Edinburgh University Press.

Chaumier, J. (1972). "Modern Methods of Documentation Offered to Marketing", *Industrial Marketing Management*, 1, No. 3.

Chorus, C.A. (1972). "The International Society for Technology Assessment", Pamphlet issued by I.S.T.A., The Hague.

de Jouvenal, (1963). *Futuribles: Studies in Conjecture*, Vol. 1, Geneva.

de Jouvenal, (1967). *The Art of Conjecture*, London.

Forrester, J.W. (1958). "Industrial Dynamics: A Major Breakthrough for Decision Makers", *Harvard Bus. Rev.*, July—August.

Forrester, J.W. (1968). "Industrial Dynamics after the First Decade', *Management Science*, 14, No. 7, 398.

Jantsch, E. (1967). *Technological Forecasting in Perspective*, Organisation for Economic Co-operation and Development, Paris.

Kahn, H. and Wiener, A.J. (1967). *The Year 2000: A Framework for Speculation on the Next Thirty-Three Years*, New York: MacMillan Company, 431 pp.

Kantrowitz, A. (1972). Statement prepared for presentation to Subcommittee of Computer Services Committee on Rules and Administration, of the U.S. Senate, March 2, 1972.

Meadows, D.L. *et al.* (1972). *The Limits to Growth*, A Report of The Club of Rome's Project on the Predicament of Mankind, Washington, D.C.: Potomac Associates.

Mueller, R.K. (1970). *Risk, Survival, and Power*. New York: American Management Association, Inc.

Schlesinger, J.R. (1967), "Systems Analysis and the Political Process", *RAND Corporation Paper*, Santa Monica, June 1967. See also "Technology and the Polity", *Harvard University Program on Technology and Society, summer, 1969*.

Sterling, Claire (1972). "Club of Rome Tackles the Planet's Problematique", *Washington Post*, 2nd March 1972.

Wills, G. *et al.* (1969). *Technological Forecasting and Corporate Strategy*, London: Bradford University Press and Crosby, Lockwood & Son.

CHAPTER 2

Technology Assessment in the U.S.A.

*"In fact, I think the survival or breakdown of our western civiliza-
tion is likely to depend on how intelligently we apply its science
and technology to our human environment within the next decade.
If we can cope with the problems our unprecedented knowledge
has created, we can do so only by properly using the tools of
knowledge. No previous civilization has had either our knowledge
or our tools. It seems to me that in this fact we have remaining
some hope that we can avoid the path of breakdown that history
suggests is inevitable for every civilization."*
                              Charles A. Lindberg, April 15th, 1970

Introduction

It is not surprising that the greatest overt display of concern
about ameliorating the worst technological excesses of the market
place has arisen in the United States of America. One obvious
cause is that Americans have benefited and suffered more than
most other societies from the numerous technologies of the twen-
tieth century and they are currently still in the van of technological progress. Stemming from the concern that Americans feel
about "the deep-seated environmental, social, economic problems
which comprise the great bulk of the problems facing the nation
today", there are two major outcomes of significance to the theme
of this book.

(*1*) The effect of the National Environmental Policy Act (NEPA)
on administrative decision making and

(2) the indefatigable efforts made to establish an Office of Technology Assessment (OTA).

In this chapter the "Section 102 Environmental Impact Statements", made mandatory for Federal Projects by the National Environmental Policy Act of 1969 (P.L. 91—190) will be discussed in their guise of technology assessments, albeit of rather specialized form. Then consideration will be given to the more general Bill H.R. 10243 (and its companion S 2302) designed to establish an Office of Technology Assessment for the Congress as an aid in the identification and consideration of existing and probable impacts of technological application.

The National Environmental Policy Act and Technology Assessment

> *"There are in nature certain foundations of justice,*
> *whence all civil laws are derived but as streams."*
> Francis Bacon

At a time when there is much criticism of governments for their attitudes towards pollution control and environmental problems (see Tinker (1972) for a typical example[1]) it is a great pleasure to acknowledge the imaginative, innovatory legislation enacted by both Houses of the U.S. Congress. The National Environmental Policy Act (acronym NEPA) is extraordinary in the way it has triggered off much public involvement with environmental problems, given the public cogent information, and provided a new effective means of bringing environmental problems before the courts of law. Appreciation of the Act's provisions may be obtained from the following extracts from a House Report by Garmatz (1971), made on behalf of the Committee on Merchant Marine and Fisheries who are responsible for overseeing the administration of NEPA, which gives a concise introduction to the Act.

----

[1] Tinker writes persuasively in this genre "Some of Britain's pollution rules are better suited to an Edwardian girls' school than to an advanced industrial society. Offenders are taken quietly on one side by the prefects and ticked off for letting the side down. There is no need for prosecutions: the shame of being found out is reckoned to be punishment enough. Carefully shielded from vulgar eyes, pollution control operates behind a deliberate smokescreen of evasion and reticence."

"NEPA's policy scope has been labelled revolutionary in its intent to set the Nation on a new course of environmental management. Its policy scope has been compared to that of the Full Employment Act 1946 which was similarly designed to bring about a national consensus on new directions for future growth and development, and to establish new machinery to bring government programs into line with innovative public policy. Since its enactment there has been a strong amount of public involvement with Section 102 (2) (C) of NEPA. This section requires each agency of the Federal bureaucracy to develop adequate means of information gathering and analysis to enable it to include in every recommendation or report on proposals for legislation and other major Federal actions significantly affecting the quality of the human environment, a detailed statement by the responsible official on:

(i) the environmental impact of the proposed action,

(ii) any adverse environmental effects which cannot be avoided should the proposal be implemented,

(iii) alternatives to the proposed action,

(iv) the relationship between local short-term uses of man's environment and the maintenance and enhancement of long-term productivity, and

(v) any irreversible and irretrievable commitments of resources which would be involved in the proposed action should it be implemented.

In developing and processing such "environmental impact" statements each agency is further required (1) to consult with and obtain the comments of other Federal, State and local agencies which have jurisdiction by law or special expertise or have been authorized to develop and enforce environmental standards; (2) to utilize these statements and associated agency comments as companion documents along with other proposal documents as they pass through the existing agency processes; and (3) to make copies of the statements and comments available to the public under the Freedom of Information Act."

From the above requirements, the Council on Environmental Quality (acronym CEQ), which is directed to review and appraise Federal programmes in the light of NEPA's policy and to make recommendations to the President, inferred the concept of a *draft impact statement which must be circulated and made public for comment prior to the final agency decision* (see Crampon, 1972).

Summaries of 102-statements filed up to the 31st January 1972, listed by agency and type of project, are given in Tables 1 and 2 (taken from the Congressional Record — Extensions of Remarks, March 2nd 1972).

The tables show that in the short space of time that has elapsed since the enactment of NEPA, nearly two and a half thousand environmental impact statements have been filed with the CEQ. Naturally, the mere number of submitted statements is no guarantee of the efficacy of the Act. There is a danger that some agencies

TABLE 1
Summary of 102-statements filed with CEQ (by agency)

Agency	Draft 102s for actions on which no final 102s have yet been received	Final 102s on legislation and actions	Total actions on which final or draft 102 statements have been received
Agriculture, Dept. of	51	97	148
Appalachian Reg. Comm.	1	0	1
Atomic Energy Comm.	36	28	64
Commerce, Dept. of	1	7	8
Defense, Dept. of	3	2	5
Air Force	2	3	5
Army	6	7	13
Army Corps Engs.	182	273	455
Navy	7	4	11
Delaware River Bas. C.	3	0	3
Environmental Prot. Ag.	10	9	19
Federal Power Comm.	18	5	23
General Services Adm.	15	24	39
HEW, Department of	0	1	1
HUD, Department of	13	15	28
Interior, Dept. of	59	36	95
Interior Boundary and Water Comm. U.S. and Mexico	1	4	5
National Aeronautics and Space Admin.	14	8	22
National Science Found.	1	1	2
Office of Sci. and Tech.	0	1	1
Tennessee Valley Auth.	10	4	14
Transportation, Dept. of	846	570	1,416
Treasury, Dept. of	1	3	4
U.S. Water Res. Council	5	0	5
Veterans Admin.	1	0	1
Totals	1,286	1,102	2,388

may make cursory impact statements simply to satisfy the letter of the law. Nevertheless, if one consults Crampon (1972), it is possible to find crude statistics which indicate that there are pressures acting on agencies forcing them to grapple with the underlying realities of the act. Up to March 1972 NEPA had generated 47 district court decisions, 15 circuit court decisions, and 3 Supreme

TABLE 2
Summary of 102-statements filed with CEQ (by kind of project)

Agency's project type	Draft statements for actions on which no final statements have yet been filed	Final statements on legislations and actions	Total actions on which final or draft statements for Federal actions have been taken
AEC nuclear development	9	11	20
Aircraft, ships and vehicles	0	5	5
Airports	34	133	167
Buildings	1	5	6
Bridge permits	10	8	18
Defense systems	2	2	4
Forestry	2	4	6
Housing, urban problems, new communities	8	10	18
International boundary	4	2	6
Land acquisition, disposal	13	29	42
Mass transit	2	2	4
Mining	4	2	6
Military installations	14	4	18
Natural gas and oil			
Drilling and exploration	3	5	8
Transportation and pipeline	5	3	8
Parks, wildlife refuges, recreation	20	16	36
Pesticides, herbicides	10	11	21
Power			
Hydroelectric	19	5	24
Nuclear	26	16	42
Other	16	1	17
Transmission	7	7	14
Railroads	0	1	1
Roads	665	397	1,062
Plus roads through parks	126	27	153
Space programmes	4	4	8
Waste disposal			
Detoxification	6	2	8
Munition disposal	2	3	5
Radioactive	5	1	6
Sewage facilities	9	5	14
Solid wastes	1	0	1

TABLE 2 (continued)

Agency's project type	Draft statements for actions on which no final statements have yet been filed	Final statements on legislations and actions	Total actions on which final or draft statements for Federal actions have been taken
Water			
Beach erosion	5	20	25
Irrigation	17	9	26
Navigation	53	97	150
Municipal and Industrial			
supply	7	1	8
Permit (Refuse act)	9	1	10
Watershed and Flood	128	229	357
Weather modification	7	4	11
Research and Development	14	6	20
Miscellaneous	19	14	33
Totals	1,286	1,102	2,388

Court dissents, all involving judicial review of administrative actions.

A most significant new (13th January, 1972) appelate decison on Section 102 of NEPA has been handed down by the U.S. Court of Appeals for the District of Columbia Circuit in the case of the National Resources Defense Council versus Morton (see E. 1886, 1972).

"The case involved a proposed sale of offshore oil and gas leases by the Department of the Interior. The sale was enjoined by the lower court on the grounds that the impact statement did not adequately discuss "(iii) alternatives to the proposed action" as required by Section 102 (2) (C) of NEPA. In upholding the injuction, the Court of Appeals provided an illuminating discussion of the range of "alternatives" which this section of NEPA was designed to encompass. In particular, agencies may not limit consideration only to alternatives which could be adopted and put into effect by the official or the agency issuing the statement. All "reasonable" alternatives must be discussed, including those which depend for implementation on legislation or executive action outside of the direct control of the agency. This is because: the impact statement is not only for the exposition of the thinking of the agency, but also for the guidance of ultimate decision makers, and must provide them with the environmental effects of both the proposal and the alternatives, for their consideration along with the various other elements of the public interest. The opinion sets forth the specific alternative possibilities for meeting the energy crisis — other than offshore gas and oil drilling — which should have been discussed in the 102 Statement in this case. In response, the Department of the Interior has since withdrawn the proposed sale for further consideration."

Quite obviously NEPA has some teeth and the American courts are prepared to guard the citizen's interests in environmental matters. Two further circuit court opinions attest to this legal concern and are briefly described next.

In the case of Greene County Planning Board versus Federal Power Commission (FPC) heard before the 2nd Circuit Court on 17th January 1972, it was ruled that an agency cannot accept a utility's proposal as an environmental statement and submit it in accordance with agency rules, as the draft environmental statement which must be made available prior to the hearing. The agency must draft its own statement wholly apart from whether the issues to be canvassed at the formal hearing are fully revealed in the applicant's statements (see Crampon, 1972).

Another case, involving the construction of nuclear power plants, was the result of a decision handed down by the U.S. Circuit Court of Appeals for the District of Columbia on 23rd July 1971: Calvert Cliffs Coordinating Committee versus U.S. Atomic Energy Commission. In this case, involving charges by three environmental groups that the AEC had failed to live up to NEPA, the court agreed with every argument of the environmentalists and directed the commission to make stringent revisions in its nuclear plant licensing process (see Barfield, 1971). In this case Judge Wright's opinion states that in "uncontested hearings" the agency must make careful examination of the draft 102-statement to determine whether staff review has been adequate and "it must independently consider the final balance among conflicting values that is struck in the staff's recommendation" (see Crampon, 1972). Although I am told, and believe because of my own experience of the AEC's excellence, that AEC staff are now making impact statements "second to none", the Calvert Cliffs case is of such precursor significance of what could happen in Europe[1] that it is worth-

---

[1] *Nuisance or Nemesis?*, written by Sir Eric Ashby's Working Party on pollution and published by HMSO immediately prior to the 1972 U.N. Stockholm Conference on the Human Environment, suggests that because of the problems that could conceivably occur from reactor wastes, the U.K. Nu-

while reiterating the summary given by Barfield (1971) of this case.

"The court of appeals decision in the Calvert Cliffs case ended an 18-month argument between the environmentalists and the AEC over the agency's new responsibilities under the environmental Act. On April 2, 1970, the AEC issued its first policy statement regarding the incorporation of the new environmental Act into its licensing process.

On June 3, 1970, the AEC amended the rules to take into account the the provisions of the Water Quality Improvement Act of 1970 (84 Stat. 91), passed in April, as well as the environmental act guidelines for federal agencies published by the CEQ in May. Regarding water standards, the Commission said that it would accept the certification of state or regional water quality agencies as adequate for nuclear power plants.

On June 29, the groups who would later file suit against the Commission under the Act — Sierra Club, National Wildlife Federation and the Calvert Cliffs Coordinating Committee Inc. — suggested extensive amendments to the AEC-proposed environmental regulations. The AEC formally rejected those suggestions on August 6; four months later, on Dec. 4, it published in the Federal Register (35 Fed Reg 18469) a final version of its licensing regulations.

Calvert Cliff suits: The coalition of three environmental groups filed two suits against the Commission.

The first, instituted on Nov. 25, dealt with specific issues at the Calvert Cliffs, Md., nuclear plant which was being constructed by the Baltimore Gas and Electric Co.

The second, filed on Dec. 7, represented a general attack on four of the new AEC environmental rules for all nuclear facilities. The court consolidated the two cases and six electric utility companies led friend of the court briefs in support of the AEC and the Baltimore utility.

Decision. On July 23, 1971, the U.S. Court of Appeals for the District of Columbia handed down an opinion that supported the petitioners' challenge to the AEC regulations on all of the four points raised.

———

(footnote continued from previous page)
clear power programme should be slowed down. Despite the fact that the UKAEA has a commendable safety record.

James Schlesinger, Chairman of the U.S.A.'s AEC, has also told a "chagrined power industry"

"Those of you who regard the response to the Calvert Cliffs as indicating a climatic change in the relationship between the industry and the AEC could well be right .... the move toward greater self-reliance for the industry had a certain historic inevitability. Such a process is always painful. It is, however, necessary. One result will be that you should not expect the AEC to ignore on your behalf an indication of congressional intent or to ignore the courts. We have had a fair amount of advice on how to evade the clear mandate of the federal courts. It is advice we did not think proper to accept .... I believe that broadside diatribes against environmentalists (are) not only in bad taste but wrong."
See Hearings (75—2250, March 2, 1972), p. 112.

The court ordered the Commission to make the following changes in its environmental review procedures:

institute an independent substantive review of all environmental factors by hearing boards in uncontested as well as contested licensing actions;

undertake an evaluation of all the relevant environmental issues in connection with every nuclear power licensing action that occurred after the Act became effective on Jan. 1, 1970. (The AEC had excluded all non-radiological environmental issues from hearings for which notice had appeared in the Federal Register before March 4, 1971.);

undertake an independent assessment of certain environmental factors — the most important being water quality effects — even though other state or federal agencies had already certified that a nuclear facility had met the relevant state or federal standards;

begin an immediate environmental review and the environmental cost—benefit balancing for plants where the construction permit had been granted before the passage of the Act, but where the operating license had not yet been issued. The court also directed the Commission to consider a temporary halt in the construction of these plants, pending a full review of the Act's provisions.

In addition to these directives the court dealt summarily with the argument that a national energy crisis was imminent. The Commission had relied heavily on this argument.

The court said: "Whether or not the spectre of a national power crisis is as real as the Commission apparently believes, it must not be used to create a blackout of environmental considerations in the agency review process .... The very purpose of NEPA was to tell federal agencies that environmental protection is as much part of their responsibility as is protection and promotion of the industries they operate'."

The reaction of most Europeans to NEPA, after hearing about the large number of 102-impact statements which have been made and the surprisingly gratifying (for the environmentalist) outcome of some of the legal cases arising, is one of surprise at the overt nature of the processes springing from the Act. When the availability of environmental information in Europe is compared with that in the U.S.A., the outcome, despite some of the obvious disadvantages of an overt process, is unfavourable to the European system where the quality of the environment appears to have become the "private concern of government and industry to be bargained away behind closed doors". Tinker (1972), the environmental consultant to the New Scientist, claims that "nanny does not always know best" and in the light shed by NEPA he could well be right.

Consider, for example, the ease with which 102-environmental statements are publicly obtained in the U.S.A.

(1) Each statement is assigned an order number which appears in the 102 Monitor (a freely obtainable publication which contains summaries of each statement) and also in the NTIS half-monthly

Announcement Series No. 68, "Environmental Pollution and Control". (An annual subscription costs $5 and can be ordered from the NTIS, U.S. Department of Commerce, Springfield, Virginia 22151).

(2) Final statements are available in microfiche as well as paper copy. A paper copy of any statement can be obtained by writing to NTIS, at the above address, and enclosing $3 and the order number. A microfiche costs $0.95.

(3) NTIS is also offering a special package in which the subscriber receives all statements in microfiche for $0.35 per statement.

The above information was easily obtained from Congressional Record — Extensions of Remarks, March 2 1972, E 1892, and the same page describes where to find in the Congressional Record out-of-print back issues of the 102 Monitor, and the 102-Impact Statements received by the CEQ between 1st and 31st January 1972.

Although NEPA is restricted to agency projects, the availability of information under the Act could well act as a model for Britain's CBI and Government *if* they decide to satisfy the recommendation made by the recent Royal Commission on Environmental Pollution (Cmnd 4894) which said that they ought to get together to work out ways of reducing unnecessary secrecy over pollution control.

Now is the time for the Rt. Hon. Peter Walker, Secretary of State for the Environment, to live up to his claim (see Walker, 1971) that "we, in my own country, are trying to develop this total approach to our environmental problems with the considerable powers that this gives to us and we welcome working with our friends throughout the world to see if there is a total international approach to these problems which is practical, speedy, and effective".

It was stated earlier that 102-statements are a form of technology assessment and if the reader will compare point-by-point the Huddle (1972a) definition of technology assessment (in the Preface to this book) and the requirements for a 102-statement it will be seen that this assertion is substantially correct. Perhaps the only real difference between the two definitions is that the 102-statements are concerned mainly with the impact of technology on the natural environment and not primarily on social and economic environment, although a generous reading and interpretation of NEPA would include the latter impacts. NEPA could thus be claimed, with some justification, to be a pioneering example of the efficacious powers of technology assessment. European countries have both kinds of assessment activity but, unfortunately, they are usually housed within the executive agencies of government. We must therefore conclude that the New World is showing the Old World how democracy should operate in this instance. Table 1 shows that even the Department of Defense (DOD) in the U.S.A. has become subject to NEPA and democracy can stretch no further than that at present. Admittedly, of the total number of 102-statements made by DOD, 455 have been made by the Army Corps of Engineers who are America's most prolific civil engineers. But this still leaves 29 which have been made by others more directly concerned with weapon systems.

One good example of a statement made by the military is the Project Sanguine Draft Environmental Impact Statement which deals with the proposed land-based terminal of an Extremely Low Frequency communication system which could be used to propagate messages to distant submerged submarines. Since such a system requires frequencies between 20 and 100 Hz and antennas of lengths around 150 miles, there is some concern that significant biological and ecological effects might be produced. Exploratory biological research has been done to see what biological reactions there are likely to be from exposure to extra low frequency (ELF) electromagnetic fields. Experiments have been carried out at higher power levels than those expected from Sanguine. The results of these experiments are claimed, in a draft 102-statement, to be

(a) ELF electromagnetic fields can be produced in the laboratory to conduct biological research.

(b) The results from exposing seeds to ELF fields during germination and early growth were indeterminate. Two of three species showed no apparent effect, while one showed evidence of retardation but no dose-related response. Additional experimentation is planned.

(c) Results from a fruit fly mutagenesis pilot study were inconclusive. Additional experimentation is planned.

(d) Reactions not related at this time to ELF exposure or other causes were observed in a bacterial mutagenesis study. Several potential sources other than ELF exposure have been hypothesized but no other positive identification has been made.

(e) Results from onion-root cytogenetic study indicated that ELF field exposure is an unlikely environmental stress.

(f) Rat fertility showed no effects from ELF exposure.

(*g*) Most of the clinical parameters studied during canine physiological experiments produced no unusual effects relatable to ELF exposure. Inconsistent changes in body temperatures and blood pressures were observed in control subjects as well as in exposed subjects. Additional studies are being made.

(*h*) No evidence of ELF effects on learning ability was observed while conducting rat avoidance learning studies.

(*i*) Perception and preference tests were performed on many species of aquatic and terrestrial animals. Test results indicated that the various biota neither preferred nor avoided areas of electric field intensities much higher than those required for conceptual ELF communications systems.

The total probable impact of the proposed project Sanguine on the environment is presently being studied in the Project Sanguine Environmental Compatibility Assurance Program. Three broad areas of study are being attempted.

(*1*) Biological and ecological assurance.

(*2*) Facilities construction and operation assurance.

(*3*) Interference mitigation.

(I have used the original 102-statement nomenclature in the above. Items (*1*) and (*2*) should be self-explanatory; item (*3*) refers to the interference generated electrically.) When all studies are complete and combined with an explicit list of alternatives to the proposed action, the environmentalists will have an opportunity to challenge the findings of intermediate and final 102-statements. If the DOD's final impact statement is accepted, the pacifist component of the environmentalists-in-opposition will only have recourse to calling for a 102-impact statement on defence and war, in general, and this will not be practical politics. If the environmentalists (pacifists and otherwise) in opposition to the final 102-impact statement succeed in making their criticism so cogently that the DOD suspends the project, they will have helped made good the affirmation made by the Department of Defense in November, 1969 that

> "Under no circumstances would Sanguine, or any subsequently developed ELF system be built .... unless it could be built in a manner entirely compatible with its surroundings."

As an aficionado of 102-impact statements, with an interest in undersea civil technology, I await the outcome with great interest.

For a slight diversion from environmental impact statements, but still sticking to the general theme of technology assessment of military projects, it is useful to insert a statement made by Senator Edward M. Kennedy, when he was giving testimony in favour of the Technological Assessment Bill (H.R. 10243), before the Sub-Committee on Computer Services of the Committee on the Rules and Administration of the U.S. Senate, 2nd March 1972. The Senator said

"The ABM (anti-ballistic missile) debate is another one which would have profited considerably from an Office of Technology Assessment. At the start of the debate, there was a paucity of information available to Congress and the public, other than the Administration view on the issue. Accordingly, I stimulated a group of scientists and scholars to come together and produce a book on the ABM which would inform the public on the issue and provide Congress with another source of expertise with which to evaluate the Administration's proposals. This effort, in effect, constituted a major technology assessment, and it convinced me of the tremendous importance and difficulty of carrying out such an analysis effectively. We cannot afford to depend upon ad hoc assessments of this sort in the future. Congress needs a strong capability for performing these 'assessments on a continuing timely basis."[1]

Typical 102-statements submitted on military matters in the month of January 1972 are given below to re-emphasise the general point that military projects are not exempt from the requirements of NEPA.

Department of Defense 102-statements, January 1972

Airforce	Commercial incineration of defoliant herbicide in Texas and Illinois
Navy	Reallocation of target facilities to Cross Cay, Puerto Rico
Corps of Engineers	Some 26 civil engineering projects

(The U.S.A. is indeed a country of strong contrasts when the

---

[1] A report by the Operations Research Society of America, which includes a detailed critical analysis of the Safeguard ABM debate of 1969, was extensively reported by both the Washington Post and the New York Times and these reports are published in the Congressional Record, Senate, S 15720, October 4th, 1971. Appendix III of the ORSA's report is also included in the same Record, S 15722. Careful scrutiny of these publications lends much substance to Senator Kennedy's remarks quoted above.

public can gain access to 102-statements on, and contest, military projects during an unpopular Asian war!)

Rather than review the bulk of the above 102-statements, it may be more informative to give some indication of the effect that NEPA has had on the Corps of Engineers, the major military producer of impact statements. The following indicative summary of the Corps' present position is mainly taken from the informative article by Wilford (1972)

> "'fifteen of the embattled Corps' major untertakings — worth several billion dollars in planned expenditures — are now bogged down in lawsuits and controversy, some before the bulldozers could start, others in midditch. And the Corps, while its basic philosophy has not changed, is beginning to make accommodations, recommending against proposals on environmental grounds, seeking out environmental authorities' advice, trying to get more public participation in planning.' Opposition to some of the Corps' major works has become a serious matter and the engineers have been 'frustrated, for example, in their plans to cut a $210 million barge canal through Florida; to dam the Delaware and a wild river in Arkansas; to dig a $387 million water-way from the Tennessee River to the Gulf of Mexico; and to flood a prairie park in Illinois and part of the Rappahanock valley in Virginia'. This naturally is somewhat exasperating for the Corps which can correctly claim to have done much for the public benefit and "lays claim to an early role in protecting the environment, especially in the West where it staked out the areas that are now Yellowstone, Yosemite, and Sequoia parks'. The leadership of environmentalists, making cases against the Corps, comes from the Environmental Defense Fund, a four-year-old organization of young lawyers with a reputation as the 'legal action' arm of the environmentalists' lobby. Their principal weapon is NEPA. 'Corps officials point out that this has created a situation in which a project is judged not on its merits but only on the question of whether its 102-statement is accurate'[1]. Uncertainty about how much detail to put into a statement is said to have 'put the Corps in a turmoil'. This is well illustrated by the success of the Committee for Leaving the Environment of America Natural (acronym CLEAN) in persuading a Federal judge to issue a preliminary injunction which effectively halted work on the Tennessee—Tombigbee Waterway because the Corps had not explained in its 102-statement what it would do with the earth excavated — 260 million cubic yards, twice as much as the amount excavated from the Panama Canal! Since the Corps is an acclaimed pioneer of benefit—cost and feasibility analysis, criticism of its 102-statements must be a bitter pill to swallow. Indeed, its critics are becoming bolder. One retired Army Colonel is

---

[1] If the 102-statement is not "accurate", I should like to know how people, outside privileged circles, can possibly know what the environmental benefits and disbenefits of any new project are. When some environmental change can be virtually irreversible, within a lifetime, the affected public deserves more than paternalistic reassurance.

quoted as saying 'Its nothing but a ritual. They come down the aisle swinging their incense and chanting benefit—cost. You can adjust the B—C ratio to justify any project. I did it myself a few times'. To counteract such statements General Clarke has insisted that 'the Corps cost—benefit estimates are conservative'."

The European outsider reading accounts like the above must wonder what the outcome of such vigorous interaction between environmentalists and "authority" will lead to. In the specific case of the Corps of Engineers (and other cases where the "official" side have amassed analytical expertise) I expect a reaction from those challenged which will involve more overt activity in benefit—cost analysis, particularly attempts at forcing the quantification of some of the economic and social benefits (and disbenefits) which are only qualitatively inferred at present.

Europeans cannot read with too much equanimity of American cases of constructive conflict in the environmental arena because, even if one discounts the imminent repetition of such actively contested debates in Europe, other nations' American interests can be involved. European companies trading in the U.S.A. can suffer. Take the Trans-Alaska Pipeline 102-statement as one cogent example in which European interests are involved. Under NEPA, the Department of the Interior's 102-statement *will* have some influence upon the commercial viability of European-owned oil fields on the Alaskan North Slope. It has already happened (see Congressional Record — Senate, March 8, 1971, S2576). The Department of Defense has commented on the Department of the Interior's draft environmental statement in such a way as to set back the Alaska pipeline project. European readers can balance possible economic loss to Europe against what they think the acceptability is of the DOD's criticism by reading the following abstract.

"(1) Expand the scope of the statement to include secondary actions associated with and stemming from the proposed pipeline such as ocean terminals in the contiguous U.S., the potential for additional Alaskan pipeline and facilities, and offshore leasing and drilling.

(2) To reduce any possible function that might arise as to responsibility, a new section should be added that delineates the jurisdiction and requisite action of the State of Alaska and each Federal agency as they pertain to the proposed project.

(3) A shorter document, with perhaps more material in technical appendices and a glossary of terms would improve readability and comprehension.

(4) It should be recognized that successful execution of the immensely complex and diverse responsibilities placed on the Authorized Officer will be extremely difficult.

29

(5) It is recommended that the proposed U.S. Labor Department Safety Standards, under the Construction Safety Act of 1969 as finally adopted, be utilized and enforcement of these standards be designated the responsibility of the Authorized Officer.

(6) The impact statement should state that the specific stipulations applicable to Military Real Estate would be obtained in a permit authorizing construction on lands controlled by the DOD. These stipulations would contain provisions peculiar to military operations as well as all of the Interior Department's environmental stipulations which would not conflict with the military missions.

(7) The statement should note that Department of the Army (DA) permits will be required for all work, discharges, and deposits in navigable waters of the United States which involve structures, activities or alterations covered in Sections 9 and 10 of the Rivers and Harbours Act of 1899, ... In addition to the pipeline itself this would include removal of material from the bed of any navigable waterway and the required port facilities constructed below high water (level). In issuing permits in the above situations the DA may impose additional provisions as appear necessary.

(8) In regard to both (6) and (7) above, acceptable design, analysis, plans and specifications, construction procedures, sequence and timing, quality controls, and other supporting data would be required by the DA for permit processing.

(9) The stipulations contained in this Impact Statement provide that the issuance of a permit for construction will be subject to valid existing rights already granted to others. ....

(10) The question of the future designation of streams and rivers as navigable within the meaning of the Rivers and Harbours Act of 1899 and the future public work projects on public lands is not specifically treated in the current stipulations. To avoid possible misunderstanding, we recommend that the stipulations be expanded to include the following:

(a) Reallocation or modification of the pipeline necessitated by the future designation of streams, rivers and lakes as 'Navigable' within the meaning of the Rivers and Harbour Board Act..... shall be at the sole expense of the Permittee.

(b) Reallocation or modification of the pipeline which might be required in planned or future public works projects shall be at the sole expense of the Permittee providing any such projects are constructed partially or fully with Federal funds or grants.

(11) In our opinion, the technical and environmental stipulations presently contained in the statement while well done, are too general to support the positive assurances given throughout the report that adverse ecological changes and pollution potential will be minimized by these stipulations. Our experience is that such stipulations in themselves cannot guarantee structural integrity or adequate protection of environmental matters. While the stipulations do establish excellent guidelines for design, construction, and operating considerations, we believe that it will be absolutely necessary for the Government to insist upon specific and detailed design, construction, operation, maintenance, and monitoring plans supported by strict review, inspection and supervision. This data would include all features associated with the project and should be approved prior

30

to authorization for construction. Such detail would facilitate a more quantitative assessment of the proposal.

(12) We suspect serious questions will be raised regarding the discussions on the alternatives. A more definitive discussion will probably answer many of these questions.

(13) There is very little information present on the timing and sequence of governmental review and authorizations.'

Further specific comments may be gleaned from the Congressional Record and these include DOD recommendations that technical specifications for the pipeline construction (and protection) be altered; that more attention be given to the detection of small oil-leaks; that the emergency valving system be made more explicit; that both marine and terrestrial clean-up procedures be spelt out; that changes be made in the information given about permafrost and geology; that examples be provided of techniques being developed to allow unrestricted passage of wildlife; etc.....

Astute readers will quickly perceive that the NEPA provision that each agency should "consult with and obtain comments from other Federal, State and local agencies..." gives agencies, other than the recommending agency, a *positive* opportunity to assert statutory *and* assumed parochial rights. The environmentalists [1] are not the only group to use the Act to protect their own interests. Although the agencies will naturally use NEPA as effectively as they can to assert their rights, it is only reasonable to assume that as some impacts of NEPA become troublesome they will urge Congress to reduce these impacts to what the agencies think is a reasonable degree. Barfield and Corrigan (1972) have said that, as a result of representations from agencies and industries, the White House will seek to restrict the scope of environmental law. These authors have even suggested that in proposing changes to NEPA, President Nixon's Administration overlooked the spirit, as well as the letter, of that law. A draft letter bearing the printed names of the CEQ Chairman, Russell E. Train, and the Environmental Protection Agency's Administrator, William D. Ruckelshaus, is alleged to have been given to the House of Representatives' Public Works Committee on 1st February 1972. It is said that the letter proposed changes in the environmental law but was hoist by its own

---

[1] Three environmental groups won a victory in the case of the Trans-Alaska Pipeline. In the case of The Wilderness Society versus Morton, 23rd April 1971, the U.S. District Court D.C. enjoined the Secretary of the Interior from issuing a permit to construct a road across public lands in Alaska from Prudhoe Bay to the Yukon River. This ruling in favour of the Wilderness Society, Friends of the Earth and the Environmental Defense Fund was made because the Department of the Interior did not prepare a 102-impact statement concerning the road and the associated pipeline project.

31

petard because it did not comply with Section 102, of NEPA, i.e. it did not explicitly state the environmental impact of the action it proposed! Naturally, under the circumstances the Public Works Committee reacted unfavourably to the Administration's proposals.

Since then the Atomic Energy Commision's Chairman, James R. Schlesinger, has pleaded that the AEC be allowed to issue interim licenses to start up reactors on reduced power and run them on full-power during emergencies, without satisfying the requirements of NEPA. The AEC Chairman was supported in this plea by Messrs. Train and Ruckelshaus, according to Dan Greenberg writing in the New Scientist of 6th April 1972.

No doubt, as the real costs of internalizing "social costs" or delaying projects (claimed by some to be necessary for satisfying essential needs) are felt, such counter-actions against the provisions of NEPA will become common. Optimistically, we should hope that reactionary moves against NEPA do not succeed beyond a reasonable degree and that environmental decisions continue to be made in the eye of the public and not behind "smokescreens". If there are, indeed, any decisions that need to be taken quickly (for strategic, social, or economic reasons without the imagined procrastination, introduced by having to make 102-impact statements) then it will be only fair to ask the decision makers to bear direct responsibility for any disbenefits revealed by an ex post evaluation made as soon as possible after the action has been taken.

The interaction of the Law with the environment, through the instrument of the National Environmental Policy Act, could very well be written about at this point. However, the subject is more conveniently dealt with after dealing with the interaction between technologists, scientists, and politicians and the latter is a major theme underlying the next section on the Office of Technology Assessment.

The American Concept of an Office of Technology Assessment

"The experts should be on tap, not on top"
Jack Brooks (1972)

Despite the commendable efforts of Lord C.P. Snow, there is still an unacceptable dichotomy between scientists and non-scientists. Differences are nowhere more apparent than between scientists and politicians, despite the apparent convergence of their effort when they tend to support the same mission-orientated, normative goals. There are, significantly, great differences in their methods of defining and assessing the values of normative goals and communicating these differences to each other. Articulate enumeration of these differences is rare but Huddle (1971) in his magnum opus *Technical Information for Congress* [1], lists the most important differences in tendency between these two groups, as follows.

> "The vocabulary of science is elaborate and specialized but objective and factual; that of politics is more everyday and is centred on value judgements.
> The rules of science data differ from the rules of legal evidence: scientific truth is established by objective demonstration and confirmed by replication; political truth is established by concensual agreement, usually after 'adversary' contest.
> Science deals with its subject matter in mainly quantitative terms, politics in mainly qualitative terms.
> The subject matter of scientific issues is foreign to the experience of political decision makers and few politicians can accept the product of scientific analysis as unqualified guidance in making political decisions.
> Basic science is isolated from personal desires, expectations, or motivations as to what is discovered; applied science is concerned with meeting a social goal but the scientific tests of effectiveness of any particular project of applied science are objective rather than subjective. Conversely, the thrust of politics focuses on human desires, expectations, and motivations; the political test of effectiveness is mainly whether or not the social response to a project is (or likely to be) favourable."

Whilst there is obviously a tendency to quibble with the pristine definition given by Huddle, it has its uses because the continuous spectrum, which ranges from the unattainable ends of 100% scientist to 100% politician, are reasonably defined and normal behav-

---

[1] This publication contains a study made "to help Congress assure itself of the quality and the thoroughness, as well as to determine (whether) the direction and validity, of the technical testimony it receives calls for a strengthening of the resources of personnel that serve it". Huddle's work was carried out under the aegis of the Committee on Science and Astronautics, the Sub-Committee on Science, Research, and Development, and the Congressional Research Service. Although Huddle is the predominant author he was ably assisted by Miss Genevieve Knezo and other specialists. In my opinion Huddle (1971) should be required reading for all politicians and assessors of technology.

iour, characteristics of each group, will obviously tend to the appropriate endpoint. Given the acceptability of Huddle's definition of the differences in thought and mode of action of the two groups and acknowledging that scientists and politicians want to live in mutual dependence, there is obviously a premium to be gained from improving communication between them. The results of better communication are likely to be that politicians will make better decisions on problems with scientific and technological content and scientists will be better able to adapt their undoubted skills in such a way that these will be accepted as necessary for the public good. Naive though it may appear to the sophisticate, this last philosophy underpins most efforts in the United States to legislate for an Office of Technology Assessment. Naturally, the rationale behind the recently passed Bill H.R. 10243 appears more compelling when fully fleshed with persuasive rhetoric about the real handicaps the legislative arm of government has suffered in the past (through not having a source of advice free of vested interest) on matters technological and about the all-pervading dangers to which the public may be susceptible as technology exponentiates itself. An interview with the only Congressman who had registered his profession as scientist and reflection on how the ascientific politicians would cope with the unavoidably vast amount of science-orientated literature (produced by the Congressional Research Service and others) necessary for their briefing, led me to a firm conviction that the American rhetoric was, if anything, underdone [1].

Background of the Legislation Aimed at Establishing an OTA

So many people have become enamoured with, and championed the cause of, technology assessment that it is now hard to discern who were the early protagonists of the movement to found an OTA. A persual of the vast amount of American literature which has grown around the subject does, however, make one thing clear. The mainspring of action to establish an OTA for Congress was Congressman Emilio Q. Daddario who chaired the Sub-Committee on Science, Research, and Development when it was set up in

---

[1] I am by vocation a scientist and this sentence helps demonstrate the plausibility of the Huddle definition quoted earlier. The sentence was to have been deleted but is left in for the sake of objectivity.

34

1963. Kiefer (1970)[1], a scientific journalist who has painstakingly recorded the OTA cause, suggests that Daddario's interest was awakened by Charles A. Lindberg, the aviation pioneer, and Dr. Jerome Wiesner Provost of M.I.T. and scientific adviser to the late President Kennedy. The description "technology assessment" was first used in a report issued by the Sub-Committee in 1966 but it has been attributed by Kiefer to Philip B. Yeager, Counsel to the parent committee. Kiefer also quotes Daddario as saying

> "We have just reached a point where the tendency to use technology just because it is there must be limited or controlled...... Our goal is a legislative capability for policy determination in applied science and technology which will be anticipatory and adaptive rather than reactionary and symptomatic."

Daddario, as Chairman of the Sub-Committee, introduced on 7th March 1967 Bill HR 6698 "to provide a method of identifying, assessing publicizing, and dealing with the implications and effects of applied research and technology" by establishing a Technology Assessment Board. This Bill was "not intended as serious legislation, but rather as a focus and stimulus for discussion of the topic". The salient features of the Bill HR 6698 were (see HR Report, 1971) as follows:

> "It (was to have) created a five-member Technology Assessment Board whose members would be appointed by the President, by and with the advice consent of the Senate. It provided that (1) no member of the Board engage in any business, vocation, or employment other than serving on the Board; (2) each member to be appointed for a five year term;...."
> "It gave the Board the duty of (1) identifying the potentials of applied research and technology and promoting ways and means to accomplish their transfer into practical use and (2) identifying the undesirable by-products of such research and technology *in advance* and *informing the public* of their potential in order to eliminate or minimize them" (my emphasis). "It provided for a 12-member General Advisory Council to advise the Board and provided that the Council members be appointed by the President."

After this catalytic Bill had been fed into, but not enacted by, the political system, the Committee on Science and Astronautcs asked

---

[1] This Sub-Committee reports to the U.S. House Committee on Science and Astronautics. Daddario chaired the Sub-Committee from its inception until 1970 when he resigned his seat to run, unsuccessfully, for the governorship of Connecticut. He is now (April 1972) Senior Vice-President of Gulf and Western Engineering Company, Manchester, Connecticut but still lends his active support to international activities in technology assessment.

for four separate studies on technology assessment to back it up (see Science, 1969). During the four-year period 1967—70, the Sub-Committee "systematically explored both scope and process of technology assessment and also effective ways to institutionalize technology assessment in both Executive and Legislative branches of government" (see Knezo, 1972)[1].

In September 1967, the Sub-Committee held a seminar on technology assessment which was designed to bring together and elicit the views of academic scholars, representatives from university-based interdisciplinary science policy groups, and spokesmen from science and public advisory groups, The result was (see H.R. Report, 1971) that

> "The participants concluded, among other things, that the legislative function was one important area where scientific information could be integrated with other demands of a highly technological society in coming to judgements on how and where to deploy technology."; and
> "A careful review and study of the seminar proceedings, however, convinced the Sub-Committee of the need for extensive further inquiries into various phases of technology assessment before legislation would be warranted."

The second action gave rise to the magnum opus by Huddle 1969, 1971). The Legislative Reference Service, now the Congressional Research Service, undertook to review the manner in which Congress had been dealing with technological issues since World War II. In 1969, the first phase of the work was published and became a "landmark reference work" and was revised and reissued in 1971. Because this work will not be familiar to Europeans working outside privileged circles (although in the democratic tradition it is available to all), it is worthwhile listing the 14 case studies which describe and assess resolution of past legislative issues with technological content.

(1) AD-X2: The difficulty of proving a negative

This study concerns a "white powder, represented as beneficial to the operation and useful life of electric storage batteries".

(2) The Point IV Program. Technological transfer as the basis of aid to developing countries

This is a "case study of the decision-making process leading to

---

[1] Genevieve Knezo's Biographical Review, although expressing views which are not necessarily endorsed by her employers (Congressional Research Service), is the best available introduction I have found to the American literature on technology assessment.

congressional enactment in 1950 of the first long-range U.S. technical assistance program for the less-developed countries of the world, the so-called Point IV Program".

(3) Inclusion of the social sciences in the scope of the National Science Foundation, 1945—47: A groundwork for future partnership

This title is self explanatory but readers interested in the schism between the "hard" and "soft" sciences will find it salutary reading.

(4) Congressional response to Project Camelot

"Project Camelot was a project in applied research in the social sciences sponsored by the Department of Defense. It was designed to study the political, economic, and social preconditions of instability and potential Communist usurpation of power in several developing countries."

Europeans should really read this case study because they will not want to duplicate the mistakes of our American friends.

(5) Congressional concern with the decline and fall of Mohole

The gargantuan project Mohole was designed to drill through six miles of overburden and into the "Mohorovicic discontinuity" of the Earth's crust.

(6) The test ban treaty. A study in military and political cost effectiveness

This case study is concerned with the Limited Nuclear Test Ban Treaty and considers many aspects of the myriad interactions between weapons technology and politics. Students of the anti-ballistic missile (ABM) controversy should read this well-presented account of recent history of the nuclear arms race.

(7) Establishment of the Peace Corps

"The Peace Corps was designed to supplement ongoing technical assistance programs, particularly the Point IV Program (see Case (2)) but also giving specific attention to the cultural dimension in technical assistance."

(8) High-energy physics, an issue without a focus

This case is about Federal funding of high-energy physics in the range involving energies of hundreds and thousands of billion electron volts. Europeans who are proud of our own collaborative multinational establishment CERN should read this case and ponder on the practicality of carrying out a technology assessment of its world-renowned activities.

(9) The Office of Coal Research. The use of applied research to restore a "sick" industry

The title explains the case study. British readers (who are recovering from the U.K. Coal Strike of 1972) might like to use it for a comparative study.

(10) Congressional response to the Salk vaccine for immunization against poliomyelitis

Controversy over the introduction of the Salk vaccine is examined in this case study.

(11) The Water Pollution Control Act of 1948. The dilemma of economic compulsion versus social restraint

Despite the early date in the title, this study is very topical indeed: how to deal with the growing problem of polluted streams and other surface water.

(12) Thalidomide. The complex problem of drug control in a free market

In this case study we see how politicians struggled with problems of medical and pharmaceutical ethics after the horrific thalidomide episode.

(13) The Insecticide, Fungicide, and Rodenticide Act of 1947

This case study "examines testimony received by a congressional committee in 1946 and 1947 on pesticide regulation, to explore the reasons for, and the implications of, the lag in awareness of the adverse secondary consequences of the use of pesticides".

(14) Congressional decisions on water projects

The subject of this case study is "the development of information pertinent to decisions by the Congress as to whether to authorize particular water projects". When extrapolatory forecasts of excess of demand over 1965 supply in the U.K. are predicted as 1100 million gallons per day (see Medford, 1968), this American case study has some significance for European planners.

Part III of Huddle (1971), added prior to the issue of the revised issue, contains four case studies of technology assessments performed by sources other than Congress. The four case studies are.

(1) Herbicides, environmental health effects. Vietnam and the Geneva protocol, developments during 1970

This case study is about herbicides which have been used extensively in the war in Vietnam and the accusations that the United States has been conducting environmental warfare against civilians and the ecology of Vietnam.

(2) Fluoridation. A modern paradox in science and public policy

In this case study we read of a "phase of technology assessment in which the ultimate decision is often made by referendum to the voters of a community, yet in which the data for decision making are commonly expressed in the relative terms familiar to scientists".

(3) The supersonic transport

In this case study we have an examination of "some of the considerations governing public acceptance of the supersonic aircraft (SST) as a commercial transportation vehicle".

(4) The economic impact of electric vehicles. A scenario by Bruce C. Netschert

An assumption about the future of electric vehicles, a "scenario", is presented in this case study.

A companion study, to Huddle (1971), was undertaken by the National Academy of Sciences and prepared by an 18-member ad hoc task force under the direction of Dr. Harvey Brooks[1]. The National Academy of Sciences' Committee on Science and Public Policy examined the difficulties of actually carrying out technology assessments, the possibilities of organizing governmental processes of technology assessment, and generally addressed themselves to the politico-legal and methodological aspects of technology assessment. Their recommendations were that technology assessment functions be instituted quickly in strategic governmental and industrial institutions.

Dr. Chauncey Starr of UCLA chaired another parallel committee on Public Engineering Policy for the National Academy of Engineering. This committee, after perusing technology assessments that had been made on educational technology, aircraft noise, and health screening presented further recommendations on the feasibility of technology assessment.

The completion and result of the above actions convinced the Sub-Committee on Science, Research, and Development that it now had sufficient information to consider trying to institute an Office of Technology Assessment for Congress. Accordingly, in the autumn of 1969 it held hearings on the institutionalization of assessment (see Hearings, 1969). At the conclusion of the hearings, the Sub-Committee with assistance from staff formulated Bill H.R. 17046 to establish an Office of Technology Assessment within and responsible to the legislative branch of government. This Bill, see H.R. Report (1971)

---

[1] Dean Harvey Brooks is Chairman of Faculty Committee, Programme on Technology and Society, Harvard University, Chairman of the National Academy of Sciences Committee on Science and Public Policy, and was also Chairman of OECD's ad hoc Committee on New Concepts of Science Policy (see OECD, 1971).

"provided that the office would consist of a technology assessment board to formulate policy and a Director to carry out such policies and administer operations of the Office. Provided that the duties of the Office would be as follows:

(1) Identify existing or probable impacts of technology or technological programs,

(2) Where possible establish cause and effect relationships,

(3) determine alternative technological programs for achieving requisite goals,

(4) make estimates and comparisons of the impacts of alternative methods and programs,

(5) present findings of completed analyses to the appropriate legislative authorities,

(6) identify areas where additional research or data collection are required to provide adequate support for the assessments and estimates mentioned, and

(7) undertake such additional associated tasks as the appropriate authorities may direct".

The Board was to be comprised of 13 members as follows:
two Senators,
two Representatives,
the Comptroller General,
the Director of the Legislative Reference Service of the Library of Congress, and seven members of the public appointed by the President

and the Director of the OTA was to be appointed by the Board to serve a term of six years. Provision was also made for special utilization of existing, and if necessary extended, services of the Legislative Reference Service of the Library of Congress and special coordination and liaison with the National Science Foundation. The General Accounting Office was also given the obligation and duties of providing financing and administrative services to the OTA. Bill H.R. 17046 also authorized $5 million for the initial establishment of the OTA for the fiscal year ending 30th June 1971 and thereafter such sums as may be necessary.

The activities of the OTA under H.R. 17046 were to be very much in the public eye. A provision that the OTA "shall submit to the Congress an annual report which shall, among other things, evaluate the existing state of the art with regard to technology assessment techniques and forecast, insofar as may be feasible, technological areas requiring further attention" ensured that the public should be kept informed. Nevertheless, to European eyes, the provisions of Section 6 (d) were the most innovatory. These were that "the Office or (on the authorization of the Office) any of its duly constituted officers may, for the purpose of carrying out the provisions of this Act, hold such hearings, take such testimony, and sit and act at such times and places as the Office deems advisable. For this purpose the Office is authorized to require the attendance of such persons and the production of such books, records, documents, or data, by subpena or otherwise, and to take such testimony and records, as it deems necessary. Subpenas may be issued by the Director or by any person designated by him. If compliance with the subpena by the person to whom it is issued or upon whom it is served would (in the person's judgement) require the disclosure of trade secrets or other commercial, financial or proprietary information which is privileged or confidential, or constitute a clearly unwarranted invasion of privacy, such person may petition the United States district court for the district in which he resides or has his principal place of business, or in which the books, records, documents, or data involved are situated and such court (after inspecting such books, records, documents, or data in camera) may excise and release from the subpena any portion thereof which it determines would require any such disclosure or constitute such invasion".

Hearings were held on H.R. 17046 in May and June of 1970 and the Sub-Committee subsequently drafted a new bill, H.R. 18469, based on these hearings which had the unanimous approval of its members. The full committee ordered the Bill reported in August 1970 and it was co-sponsored by the Chairman of the Full Committee and all Sub-Committee members. A companion bill from the Senate was also introduced and referred to the Committee on Rules and Administration but no action was taken in Committee or on the floor of the Senate on this proposal. Nevertheless, the text of H.R. 18469 was offered on 16th September 1970 as an amendment to the 1970 Legislative Re-organisation Act (84 Stat 1140) but was ruled not germane on a point of order.

Representative John W. Davis of Georgia (Daddario's able successor, Chairman of the Sub-Committee) reintroduced the bill as H.R. 10243 in February 1971 and the Full Committee reported the Bill in August 1971. (In July 1971 a companion bill, S 2302, had been introduced and referred to the Committee on Rules and Administration by Senators B. Everett Jordan[1] the late Winston L. Prouty, and three other co-sponsors). Bill H.R. 10243 passed the House, with amendments, on the 8th February 1972, by a 256 to 118 roll-call vote (see H.R. REP 92—865, 1972). The major amendment was offered by Representative Jack Brooks of Texas.

---

[1] Senator Jordan is Chairman of the Committee on Rules and Administration, at the time of writing (April, 1972).

Though the Senate Bill S2303 specifically agreed with the House Bill 10243 (as reported out) on the recommended membership of the Technology Assessment Board, Mr. Brooks successfully carried an amendment that the membership be reconstructed to consist of five Senators and five Representatives. Three of each five to be from the majority party and two from the minority party. Furthermore, Mr. Brooks was successful in removing the subpena and other powers given to the Director of the OTA in H.R. 10243 and 17046. Brooks told his House colleagues

"We are trying to create an agency to give Congress technical information, so if we must create one let us create one that we control"; and that his amendment would

"take away from the Director of the Office the power to initiate assessments and run the whole business in the way he pleases. This director now has subpena power and the right to call people and set hearings and initiate hearings and report findings. Instead my amendment provides that the initiation of the work will be by the congressional committees and the OTA Board. My theory is simply that experts are to be employed by a committee and utilized by a committee to give their advice and listen to it and appreciate it and make the decision yourself.".…. "I say that we would be wise if we are going to spend the money to have 50 or 60 or 70 experts that we, the Congress of the United States, are going to name them".

Mr. Davis and his Sub-Committee had considered the type of argument put forward to support the Brooks' amendment and Rep. Davis made the following remarks in opposition.

"We considered this alternative at length and we concluded that it would be an inadequate answer and that the Board needed a mixed and more balanced representation.

In the first place, I do not think there are many Members of Congress who could afford the time that ought to be given as a member of the Board. You will notice that, among other things, the Board would be required to submit an annual report, which I think will have a big impact on many government agencies.

The purpose of the committee bill is to set up as good a board as we know how to set up. We decided that there ought to be two Members of the House of Representatives on the Board and two members of the other body on the Board. We thought that the head of the Congressional research Service of the Library of Congress ought to be on the Board and we also thought that the Comptroller General ought to be a member.

We then determined that there shall be four very highly qualified public members. We talked about that matter at great length and the only person that we could think of who would be appropriate to name those four public members was the President of the United States. The President did not ask to be named. So far as I know, the Executive has had a completely 'hands off' attitude towards the whole bill, this whole piece of legislation, apparently viewing it as a matter for action by Congress, since it is an arm of Congress. This office of Technology Assessment is not even empowered to make recommendations. That is beyond the scope of its power. The only thing it is empowered to do is to develop information, to evaluate the information that it develops. That is the intent of that. If we were placing some of our power in somebody else's hands, that would be a horse of another color. But we are not doing it and I think in this instance my friend from Texas has found he is on a witch hunt. He has found a danger that just does not exist. We think we have set up a Board that will be a good Board. We think that the four public members will count it a very high honor to serve on the Board.

We have set it up so that the Director of the Board will be a well-paid man. He will enjoy a status as high as that of the head of the National Science Foundation. If the Board is really a well-operating Board, if it establishes for itself a high reputation, it will be a great force and Congress will look to it with confidence. As we have said so often here this afternoon on the floor, I think it will save our taxpayers millions upon millions of dollars if the Office of Technology Assessment fulfills the function we hope to give it today. For this reasons, Mr. Chairman, I ask that the amendment be voted down."

Despite these remarks in opposition to the Brooks amendment it was carried by 29 ayes against 19 noes. The full texts of the Bill H.R. 10243 (as amended and passed by the House) and Bill S2302 are given in the Appendix. The importance of the passing of Bill H.R. 10243 may be gauged from two complimentary statements made in the House before the roll-call vote.

"The committee report on this important legislation shows 39 congressional committees or sub-committees plus four joint committees or subcommittees whose activities and needs are directly applicable to technology assessment. The Congress was at a similar informational watershed when the Congress inaugurated the Legislative Reference Service in 1915 and the Government Accounting Office in 1921. The Office of Technology Assessment authorized in the bill will be fully responsible to the Legislative branch and would be an important conduit for information regarding the state of our technology and its support, management, or control."

Representative Fuqua

"If this new office is approved it would be the first such authorization toward the creation of a new mechanism to aid the Congress since the GAO (General Accounting Office) was founded in 1921. It would also be the first action to create a new information aid for the Congress since the Legislative Service was formed in 1951."

Representative Esch

Up to June 1972 the OTA legislation was being actively pursued. However, at the time of concluding the writing of this section we are in a U.S.A. election year and the OTA legislation is languishing due to a variety of reasons, e.g. Senator Jordan's primary election struggle and some discussion on the Brooks amendment to H.R. 10243. This loss of momentum will, no doubt, be corrected after present political exigencies are over. I confidently expect the OTA concept to be implemented in the U.S.A.

See note added in proof, page 55.

The Law and Technology Assessment in the U.S.A.

In the last two sections of this Chapter, I have attempted to report two conceptually related developments of legislation for technology assessment in the U.S.A. The first dealt with the National Environmental Policy Act (NEPA) and the way the Act's provision for 102-impact statements has boosted one specialized form of technology assessment (concerned with environmental problems). The second described the inexorable[1] progress Americans have made in their attempt to provide the legislative arm of government with technology assessment support. Both of these developments should have raised questions in the reader's mind about the interaction between the legal system and the assessors, their assessments, and the technological problems which should be, or have been, assessed. The need for technology assessment is almost irrefutable, yet legally the concept raises numerous queries. For example, Representative Hanna of California expressed concern, in the House debate on the establishment of the OTA, when he said

"It does strike me as rather singularly important that this body (the House of Representatives) should answer the question to the people of the United States: Who is in charge? Everywhere we look we see that the Congress is reacting to conditions that have been established by the decision makers whom the people of the United States have not put in charge. They are the people who are pursuing activities in this country the results of which become extremely burdensome to the American people.

Those persons do not have any constituency to whom they must respond. They are simply going to take advantage of all the opportunities to carry out to the greatest extreme the activity to which they have

---

[1] As a European assessor of technology I am full of admiration for the efforts made in the U.S.A. to establish a O.T.A. and adjectives like inexorable, indefatigable, and commendable all naturally spring to my mind. However, some American writers use unkind, pithier, comment: for example, "miracle of legislative revival from the dead", "legislative Lazarus" (see Science, 1972).

44

addressed their time, talent and money. Yet inevitably the results of that activity create in our society conditions which are then brought to Congress and we are asked to do something about it. By that time the load is so great it would give the Jolly Green Giant a double hernia and the costs of doing something about it are so great we find ourselves facing deficits the like of which even some of our most concerned gentlemen in this House would never have dreamed possible we would have to face in this country and in this Congress."

Apart from the rather obvious use of the legal system in enforcing some of the provisions of NEPA there are deeper issues at stake and the concerns of laymen, not always allowed the same effective platform as Mr. Hanna, are eventually beginning to percolate into the consciousness of the work-a-day lawyers. Distinguished lawyers have naturally mirrored the onset of this consciousness. They have launched an extensive body of new literature surveying the relationship between technology and the law (see Knezo (1972) for a bibliographic review of these publica- some of which are discussed below).

Dr. Milton Katz, Director of International Legal Studies at the Law School of Harvard University, helped pioneer the present legal involvement with technology assessment when he appeared as a witness before the November and December 1969 Sub-Committee hearings on technology assessment (see Hearings, 1969). He considered that the processes of technology assessment were pervasive and diffused throughout society in both public and private sectors.

"In the private sector, technology assessment occurs whenever a business enterprise contemplates an investment that would introduce a new technology or expand or modify its use of an existing technology. In the larger industrial enterprise, technology assessment takes place in connection with research and development, especially at the point of decision whether to move a prospective or modified technology from the laboratory into production. By extension, technology assessment also takes place in investment companies and banks to which business enterprises turn for capital or commercial financing. Assessment in these situations takes the familiar form of estimates of anticipated income and expense or, to use a broader terminology, cost—benefit calculations.

I would like to emphasize one point about the assessors who are, in fact, the whole business community. Whether or not they are aware of the fact, the existing legal order infuses their calculations. It is the legal order that determines which of the anticipated costs and benefits are taken into account by the enterprise and which are ignored.

An electric light company, for example, which contemplates the installation of a new powerplant will treat the fuel to be consumed as a cost, but not the smoke that may pollute the surrounding air nor any of the

45

waste products that may be discharged into nearby streams. The company's management may anticipate a public relations problem from the pollution. If you look hard at what they mean by a public relations problem, it appears that they have in mind a risk that aroused public opinion may generate changes in some applicable aspect of the legal system.

In the ordinary course of business, the company will calculate the estimated costs and benefits of the prospective installation without reference to any damage to the community caused by the smoke or other waste products. In the language of the economists, pollution of the community's air or streams through the operation of the powerplant would be a 'social cost', not a cost of the enterprise; it would be an 'external' and not an 'internal' cost. The economic mode of analysis is an indispensable tool for technology assessment. But I want to emphasize that the economic analysis takes for granted a particular posture of the existing legal system.......

It is a 'social' and 'external' cost only if and to the extent that the legal system happens so to decree. The legal system can alter the incidence of a cost by recognizing a cause of action in tort against the company. A judgement in tort will transfer back to the company the cost previously suffered by the community..."

Katz (1969), in a paper written for lawyers, examined "... the implications of certain doctrines and theories of tort liability for technology assessment". His conclusions were that the doctrines have helped private actions and internalized the social costs within the enterprises and so spurred "the enterprises to apply the resources of technology and management to reduce the costs". Working for the Harvard University Program on Technology and Society, Dr. Katz after completing a review of recent court cases suggested that the tendency to internalize social costs be encouraged as beneficial to both technology assessment and tort law. He also thought that the "cost burden should be spread widely enough, however, to avoid penalizing any single enterprise that might become the target of a lawsuit or the victim of an accident. It should not be spread so widely, however, as to nullify the incentive to all firms within an industry to make more sensitive account of harmful side effects and make more determined effort to eliminate them. The accent should be kept on therapeutic deterrence and incentives, along with compensation for the victims".

The reader will, of course, realize that any attempt at "spreading" of costs will no doubt lead to other interwoven legal problems, which could be profitable business for lawyers.

Another active examiner of the legal problems of technology assessment is Harold P. Green, National Law Center, George Washington University. He claims that lawyers have been waiting in the wings, hesistant about contributing to the debate, because the

46

enactment of legislation introduces unfamiliar issues and science. The apparent transfer of the protective role from the courts to the legislature is said to have profound significance for lawyers and Green (1967) believes that the legal profession, therefore, should consider its specific role in assessing social consequences of the advance of technology. In Green (1970) criticism of the Bill to establish an Office and Board of Technology Assessment is made.

"The Board's findings, conclusions, and recommendations ex cathedra will not reflect what the public wants and what the public fears. The very expertness and authority of the Board will detract from public discussion and debate .... Congress and the public will still be faced with a barrage of propaganda from the technology's proponents and we may be certain that they will attack and seek to minimize the negative considerations brought to light in the course of these assessment processes.

These considerations lead to the conclusion that what is needed for the technology assessment function is an agency which would act as a responsible devil's advocate or technological ombudsman and play the role of adversary in the Congressional and public forums. ....

In discussing the manner in which such an agency should operate, it is necessary to distinguish between two classes of technology. On the one hand, there is technology which is essentially private .... primarily as a consequence of private, profit-seeking investment. On the other hand, there is another class of technology which is developed primarily as a consequence of government investment.

In the first of these cases, the market place operates as a continuous technology assessment mechanism .... Both the sponsors of the technology and the customer—users also assess the potential costs to them which may result from liability to others arising out of such hazards. Liability itself results from an assessment process by the courts since the social utility of the activity producing the injury to the plaintiff is frequently weighed (sometimes explicitly) against the deterrent effects of the imposing liability ...... The second case, that of government-sponsored technology, is quite different. .... the 'deep pocket' of the government supports technology development merely because desirable benefits are foreseen even though there are no market incentives, while none of the restraints and deterrents which are present with respect to privately developed technologies are operative in this case .... Not infrequently .... in the case of government-sponsored technologies, it is difficult for the opponents to obtain relevant information.......

Frequently the establishment seeks to discredit the opposition ad hominem and this exacerbates the situation forcing the opposition to take an extreme position which makes it easier for the proponents to discredit their contentions on their merits..........

I believe the only effective mechanism for protecting the public against the onslaught of new government-sponsored technologies is the type of devil's advocate mechanism I have proposed."

In advocating the importance of an adversary process "in order to assure that the assessment function will properly and effectively protect the public interest", Dr. Green has proposed an "ombudsman" mechanism which has something in common, if not in inspiration, with Dr. Kantrowitz's advocacy technique for "increasing the presumptive validity of statements of fact" (see Chapter 1). Notwithstanding the detailed differences between the proposed advocacy mechanism, such suggestions are useful because anything which enhances fair juxtaposition of reasoned advocacy has much to commend it. Yet I find Dr. Green's belief in the efficacy of the market place as "a continuous technology assessment mechanism" rather ingenuous and suspect that *both* classes of technology defined by him in the above quotation require equal surveillance and criticism.

American lawyers have not only commented individually on technology assessment, they have also worked in concerted ways. For example, the Denver Law Journal's "Conference on the Implications of Science-Technology for the Legal Process" (see Young, 1971), encouraged several articles dealing with the interaction between technology assessment and the law. Papers presented at the Conference included

"Science and technology versus law, or a plague on both of your houses", by Wilbert E. Moore,

"The social control of science and technology", by Michael Baram,

"The law school as a centre for policy analysis", by Arthur S. Miller,

"Political adaptation to a technology-surfeited society", by Franklin P. Huddle, and

"Protecting us from ourselves: the interaction of law and society", by James W. Curlin.

As an indication of the tenor of this conference, an abstract from one contribution is sufficient. Curlin made a similar point to Katz on social and external cost but with rather more emphasis on the consequences of not internalizing social costs. Curlin said

"Internal assessment of product suitability, even in this period of manifest social concern, emphasizes the calculus of the 'economic cost—economic benefit ratio', with little concern for the consequential damages hidden in the 'social cost—social benefit ratio'. By emphasizing profit margins and at the same time ignoring concomitant social costs, these expenses are transferred to the general public who absorb, inter alia, consequences of deficient automobile design, release of pollutants into the environment and traffic fatalities as a result of poor highway design. It is these societal costs which disrupt social order and undermine political and judicial stability.

48

Much of the furor about technology centers around the morality of permitting external costs to be transferred to society in general, without the informed consent of those who finally must bear the cost. Science—technology always has a disquieting effect on social stability. Indeed, any agent which significantly disturbs the dynamic equilibrium of society tends to create conflict. Even the undeniable benefits of technology are subversive to social stability [1].

Curlin believes that the market, courts, and legislature presently serve ad hoc roles in technology assessment. Apropos of the market, in opposition to Green, Curlin said

"The market has failed to give any indication of being a viable technology assessment mechanism .... Failure of the market in its assessment role is attributable partially to the consumer's failure to recognize potential hazards and secondary consequences and partially to an unconscious awareness that the societal costs (external diseconomies) fall upon the general public, hence do not directly affect the specific group or person who purchases the fruits of technology".

Although acknowledging that the courts act as "sentinel of technology impact", Curlin thinks that the "chronic, insidious damage emanating from modern technology mocks traditional (common-law) doctrines which require showing of harm and proximate causation". He believes that standards established by the courts presuppose that the harm has occurred and he quotes Miller (1968): "Law is the articulation of the answers of yesteryear". Despite the possibility of the legislature enacting statute law in anticipation of a technology-induced social problem, Curlin believes that the legislative process also operates after the fact. His study of the legal literature culminates in arguments in favour of new institutions (for the assessment of technology) effectively staffed with natural, social, and engineering scientists and *innovative* in their approach to assessment. (This last requirement is, in my opinion, particularly necessary. See Chapter 4, about the reaction to new methodology in the PAU)

A full European discussion of some of the legal consequences of technology assessment is advisable amongst the high technologists

---

[1] I am more and more impressed by the way the physical principle established by the great French chemist Le Chatelier (1850—1936) applies also to social and personal systems. Those who have won a pyrhic victory, for change against unreasonable authority, will testify to the truth of Le Chatelier's Principle: "If to a system in equilibrium a constraint is applied, the system tends to adjust itself so as to oppose the constraint and restore equilibrium".

and lawyers in the EEC. Readers who are interested in pursuing the American literature on this fascinating subject as preparation will find Crampon (1972) and Knezo (1972) particularly good sources of material. Literature of American thought on the interaction between the Law and Technology Assessment is very substantial and, without writing a complete book on this one facet of technology assessment, it is impossible to give more than information on, and subjective reaction to, the topic. The direction of my present thought has been inspired by a favourite quotation, from Tribe (1971)[1], which should suffice to round off this section.

"Ultimately, of course, there will be many situations for which neither actions for compensatory damage nor taxes on 'side-effects' will be appropriate means of attaining an optimal allocation of resources — as when the needed market modification is the internalization not of costs but of benefits. Thus, when the number of potential beneficiaries of a technologically related development (such as, for example, a completely degradable and non-persistent pesticide) is very large and when the problem of excluding taxpayers from the enjoyment of benefits is particularly difficult, the developer will invest a less than optimal amount unless government subsidises his undertaking. In such cases, the proper market adjustment will typically be the establishment of substantial economic rewards for the socially useful technological development, paralleling the patent system but clearly identifiable from it ....

Interestingly enough, there are reasons to believe that the United States Patent Office has occasionally refused to award patents because of the perception of undesirable side effects. I have been told, for example, that a patent related to the 'morning-after' pill was initially denied on the grounds that it would induce immoral behaviour. The obvious ineffectiveness of such a denial as a means of protecting the public, coupled with the clear impropriety of lodging such vast powers of technological censorship in a body like the Patent Office should make it plain why such a course is ill-advised.

In the converse possibility of awarding patent protecting more liberally or in a more profitable form for inventions of particularly high 'social utility', perhaps the fundamental problem apart from the growing inappropriateness of the discrete 'invention' as a subject for the invention system is that the value of a patent award is intrinsically limited by the marketability of the process patented. Such a value is negligible for inventions in areas where the economic market itself would not have stipulated significant inventive activity, precisely the areas of invention which a system of positive public incentives must be particularly designed to foster".

---

[1] Laurence H. Tribe made this comment in his paper delivered at the Aspen Conference on "Technology: Social Goals and Cultural Options", sponsored by the Aspen Institute for Humanistic Studies and the International Association for Cultural Freedom, August—September, 1970.

Now there is a positive suggestion for the legal researchers, legislative makers of statute law, and the judiciary to attack!

American Industry and the Office of Technology Assessment

"Technology assessment could be technology arrestment" is an understandable reply given by some industrialists and business men when they are asked about their reaction to the present American legislative trend. When this response is coupled with the fact that most experienced analysts know that, in the present embryonic state-of-the-art, "technology assessment's plausibility exceeds its capability" it seems as if there could be a prima facie case for not encouraging the institutionalization of assessment[1]. However, the only two opinion polls of industrial reaction, known to me, do not show active opposition to the concept of an independent Office of Technology Assessment.

Peat *et al.* have carried out a survey of Technology Assessment in the United States which is indicative of the reaction of American industrialists but this is, unfortunately, still (March, 1972) under consideration by the National Science Foundation and has not yet been published or distributed. Another survey carried out by the American journal Industrial Research, in 1970, published Sept. 1971, gave the following information.

71% of the 1,325 respondents thought that a special government board was warranted because of the undesirable side-effects of technology,

67% thought what technology assessment would not cause an appreciable "arrestment" in critical areas of R & D,

62% optimistically thought that byproducts of technology could be identified in advance,

62% thought that groups with vested interests in developments, identified as potentially harmful, desist from pursuing such developments,

70% supported Rep. Daddario's proposals for a Technology Assessment Board,

---

[1] Gabor Strasser, formerly with the President's Office of Science and Technology (now Director of Planning at the Battelle Columbus Laboratories) was so concerned with the misinterpretation of "assessments" to "arrestment" that he coined the acronym
PAYOFF for Plans and Assessments to Yield Options for the Future.
For similar reasons Huddle (1972a) suggested the following acronyms:
APT for Analysis of Policy for Technology, and
ETA for Evaluation of Technological Alternatives.

51

92% thought Congress was not well informed concerning technological advances, and

77% believed that Congressmen would only tend to use information on technology when it suited their own political purposes.

Such support for technology assessment is very gratifying but one is entitled to harbour reservations about what the reaction of these same respondents would be to an unfavourable assessment of their own specialized fields of technological activity. Even though the explicit powers of subpena have been removed by the Brooks' amendment from the Bill to establish an OTA there are still subpena powers implicit in the Congressional membership of the Board and the only proof of industrial good-will shall be favourable industrial reaction to an actively pursued assessment.

Industrialists will essentially be affected by an OTA in the sense that potential "guilt", before the technological action, will be in question. Innovation (1971), considers the internalization of social costs and asks the questions "It is more efficient or more economical to add these social costs at a different point in the development cycle? Or can one, in a sense, substitute assessment for litigation and pre-empt damage suits?". The answers to these last questions have not been objectively attempted by the industrialists because without real assessments to examine and real predictions of "threat" to evaluate, the answers can only be assertive and subjective. Dr. Leon Green, of Lockheed Aircraft Corporation, has written that

"the principal danger is not that new technology will be furtively imposed upon a blissfully ignorant public without adequate assessment, but rather that it will be over-assessed to the point of harassment by hysterical (and hardly democratic) scientific Philistines, principally from the sinister side of the political spectrum".

Science (1972) contains further reactionary remarks claimed to have been made in an interview by William O. Baker, Vice-President research and patents, Bell Laboratories. He is reported as saying

"Technology assessment can subvert the principles at the very heart of democracy."

"There is no basic or natural concordance between the capability to do science and technology and the public purpose. The efforts of making technology assessments may well destroy the long-range values of the technology itself. When you attempt to pre-judge certain alternatives, you thereby bias possible later and realistic choices of action.

Technological development flourishes only with a more delicate balance."

Further qualified support for the industrialists comes from David M. Kiefer, writing in the Wall Street Journal of January 7, 1972, he makes the following points.

"Certainly industry needs a technological stance of its own, if only for self-preservation so that it may respond to pressures either from government or from the private interest groups. Maybe what it needs is a counter commercial development staff or long-range debunking group within the overall corporate structure. Such a group would assess business objectives and priorities not merely in the customary terms of short-range profit and sales growth but also in terms of social responsibility and consequences .... The role of corporation Cassandra will hardly help anyone to win friends in the executive suite. But industry may soon find that it cannot afford, either from an economic or public relations standpoint, not to have a few Cassandras on the payroll lest it be clobbered by outside pressures.

Much of the responsibility for making assessments and putting them to use in controlling technological progress must, of course, rest with government. The function of government, after all, is to set ground rules and establish priorities within which private groups may operate.

Businessmen's fears of technology assessment are not unreasonable. By adding new uncertainties to the research and development equation that is already strewn with risks and ambiguities, assessment could well discourage private investment and undercut innovation. .... It will certainly not be easy to force businessmen to account for all the indirect consequences and spill-overs that they have long been accustomed to ignore or pass on to the public at large.

But if the scientific and industrial community remains unequipped to sort out good uses of technology from bad, all science and technology may suffer. A growing impatience with the failures of technology could turn the public against science, causing very heavy social costs. Many remedies for past failures will only be found through the introduction of still newer and more sophisticated technology."

Kiefer (1970) also quotes former Du Pont Vice President, Samuel Lenher, as giving some support to the idea of an OTA: "It's a social and technological experiment that should not be delayed but it should only be set up as an experiment at this stage, rather than as a rigidly structured new arm of government that might be very difficult to halt if the experiment failed to work as expected". In the same article Kiefer cites Dr. Morris Tanenbaum, general manager of the engineering division of Western Electric, as being convinced that some form of technology assessment mechanism is needed. But we are also told that Tanenbaum thinks that badly applied technology assessment could do more harm than good. Tanenbaum's minority appendix to the National Academy of Sciences report on technology assessment states that it "could

53

discourage private investment in areas of technology which are at the focus of assessment activity. These may be areas where innovation is most important and private participation most desirable" and he warns that the assessment process must be arranged to operate "without irreparably damaging the systems of innovation that it is designed to stimulate and guide".

Now that the first officially "blessed" attempts to perform, and politically act upon, objective technology assessments are imminent in the United States it may not be necessary to speculate much longer about where the majority industrial reaction to this new activity will fall. Europeans, with luck, should soon have some substantial evidence from across the Atlantic to guide their own approach to the knotty problem of harnessing technology for the greatest collective benefit.

Quite obviously the managerial members of industry's "technostructure" will not react in a way that will leave them open to the accusation that they are really interested in the creation of a surveillance-free, laissez faire, new "industrial state" (see Galbraith, 1967).

Neither will they be acquiescent if assessments are made by the OTA which appear unnecessarily harmful to their business interests. As in all new legislation, mistakes will be made by both OTA assessors and industrialists when trying to implement the OTA Act. One way of forging a new tool that is needed, and you do not know how to make, is to legislate for its use and then wait for it to emerge from the heated controversy which has been induced by the attempt to satisfy the legislation. I particularly admire the Americans for tackling a world-wide problem in this way; their approach reminds me of Marshall Ney's alleged instruction to the officer in charge of his execution, "Forget the formulas and get on with the job!" As a sometime purveyor of mathematical aids, I take "formulas" in the above to mean formulas of behaviour which avert, and appear to dispel, friction.

It is all too easy for Europeans to criticize the energetically pursued head-on American approach to the solution of horrendous technological problems. However, the problems are of such magnitude, and we have had such warning of them, that the handling of them behind the customary smokescreen laid by some industrialists and some governments' executive branches will no longer suffice to calm public concern. In Europe we hear that the Rt. Hon. Peter Walker admires Disraeli as "a rather forward-looking gentleman who realized the future strength of the environmental lobby"

54

when he supported legislation to set up the British Alkaline Inspectorate in 1873. If his remark was an attempt at leavening an address he made to a joint colloquium of the Senate and the House of Representatives (see Walker, 1971), his research before launching into oratory should be admired. But if it was a Tory assertion of the invariance of Conservative Party principles, we must hope that this party, or some other, will throw up a "forward-looking gentleman" more actively concerned with technology assessment before another century passes. Ex-Labour party M.P. Jeremy Bray does at least recommend that Disraeli's Alkali Inspectorate, now responsible for limited aspects of pollution control, be "reconstituted as the British Pollution Control Service responsible for inspection and enforcement of legislation and for monitoring and avoidance of hazards to the health and safety of people and to the environment". Yet Bray's suggestion for incremental improvement upon Disraeli's legislation is only partially commensurate with present British and European needs and far behind the imaginative American approach.

Should European readers consider the American legislation to be going too far and not wish to see comparable legislation in Europe, they might console themselves by remembering that it was another American who made the following appropriate remark.

"I know no method to secure the repeal of bad or obnoxious laws so effective as their stringent application."

Ulysses Simpson Grant, President,
Inaugural Address, 4th March 1869.

Note added in proof

On September 13, 1972, the Senate Committee on Rules and Administration met and unanimously voted to report favourably H.R. 10243, with an amendment in the nature of a substitute. H.R. 10243, as reported by the committee, passed the Senate on September 14, 1972, for return to the House. Senate and House conferees met on September 21, 1972, and agreed to report favourably H.R. 10243 as passed by the Senate with certain minor and technical amendments. The Senate agreed to the conference report (H. Rep. 92—1436) on September 22, 1972. The House passed the report on October 4, 1972. The Technology Assessment Act of 1972 was signed by the President on October 13, 1972, becoming *Public Law 92—484.*

See "Technology Assessment for the Congress: Staff Study of the Subcommittee on Computer Services of the Committee on Rules and Administration, United States Senate", November 1st 1972, for the first Senate publication on the subject of technology assessment and an interesting discussion on the operational problems which will arise in implementing the new OTA Act (1972).

References

Barfield, C.E. (1971). "Energy Report Calvert Cliffs Decision Requires Agencies to get Tough with Environmental Laws", *National Journal*, 18th September 1971.

Barfield, C.E. and Corrigan, R. (1972). "Environmental Report/White House Seeks to Restrict Scope of Environmental Law", *National Journal*, 26th February 1972.

Crampon, R.C. (1972). "The Effect of N.E.P.A. on Decision Making by Federal Administrative Agencies", *Administrative Conference of the United States*, 726 Jackson Place, N.W. Washington, D.C. 20506. March 2nd.

E 1886 (1972). *Congressional Record — Extensions of Remarks*, March 2nd.

Galbraith, J.K. (1967). *The New Industrial State*, Houghton Mifflin, Boston.

Garmatz, A. (1971). "Report by the Committee on Merchant Marine and Fisheries", *Administration of the National Environmental Policy Act, House Report No. 92—316*, Washington, D.C., U.S. Govt. Printing Office.

Green, H.P. (1967). "The New Technological Era: A View From the Law", *Program of Policy Studies in Science and Technology, Monograph No. 1*, The George Washington University, Washington, D.C., November 1967.

Green, H.P. (1970). "The Adversary Process in Technology Assessment" *Technology + Society* (formerly *The Technologist*) 5 No. 4, March.

Hearings (1969). "Technology Assessment", *Committee on Science, Research, and Development, Committee on Science and Astronautics, U.S. House of Representatives, 91 Congress, 2 Session*, U.S. Govt. Printing Office, Washington.

Hearings (1972). "Office of Technology Assessment for Congress", *Sub-Committee on Computer Services of the Committee on Rules and Administration United States Senate, March 2*, U.S. Govt. Printing Office, Washington.

H.R. Report (1971). "Establishing the Office of Technology Assessment and Amending the National Science Foundation Act of 1950", *House of Representatives Report No. 92—469, 92 Congress, 1st Session*, U.S. Govt. Printing Office, Washington.

H.R. Rep. 92—865 (1972). "Establishing the Office of Technology Assessment and Amending the National Science Foundation Act of 1950", Congressional Record House, U.S. Govt. Printing Office, Washington.

Huddle, F.P. (1969, 1971). *Technical Information for Congress*, Printed for the use of the Committee on Science & Astronautics April 25, 1969 and revised April 15, 1971. U.S. Govt. Printing Office, Washington.

Huddle, F.P. (1972a). *"A Short Glossary of Science Policy Terms,"* Science Policy Research Division, The Library of Congress, preliminary draft not yet published.

Huddle, F.P. (1972b). "The Social Management of Technological Consequences" *The Futurist*, VI, No. 1, February.

Innovation (1971). "Making Technology Assessable", *Innovation*, No. 25, October, 56.

Katz, M. (1969). "The function of Tort Liability in Technology Assessment", *University of Cincinnati Law Review*, 38, No. 4.

Kiefer, D.M. (1970). "Technology Assessment", *Chem. Eng. News*, 48, Oct. 5, 42—45.

Knezo, Genevieve J. (1972). "Technology Assessment: A Bibliographic Review", *ISTA Journal*, in the press.

Medford, R.D. (1969). "The Application of Technological Forecasting", in Arnfield, R.V. ed., *Technology Forecasting*, Edinburgh University Press, pp. 268—269.

Miller (1968). "Science vs. Law: Some Legal Problems Raised by 'Big Science'", *Buffalo Legal Review*, 17, 593.

OECD (1971). "Science Growth and Society. A New Perspective". *Report of the Secretory-General's Ad Hoc Group on New Concepts of Science Policy*, Paris: OECD.

Science (1969). 14th November 1969.

Science (1972). "Office of Technology Assessment: Congress Smiles, Scientists Wince", *Science*, 175, 3 March.

Tinker, J. (1972). "Britain's Environment — Nanny knows Best", *New Scientists*, 53, No. 786, 9 March.

Tribe, L.H. (1971). "Legal Frameworks for the Assessment and Control of Technology", *Aspen Conference on Technology: Social Goals and Cultural Options*, sponsored by the Aspen Institute for Humanistic Studies and the International Association for Cultural Freedom, August 29—September 3, 1970.

Walker, Rt. Hon. P. (1971). Statement made during a Joint Colloquium before the Committee on Commerce U.S. Senate and the Committee on Science and Astronautics House of Representatives, May 25 1971. Text of Walker's speech may be found in *International Environmental Science*, p. 22, U.S. Government Printing Office, Serial No. 92—13, Washington.

Wilford, J.N. (1972). "Corps of Engineers Caught Up in Battle Of the Builders Against the Preservers", *The New York Times*, Sunday, February 20th, 1972.

Young, D.R. (1970). *Conference on the Implications of Science—Technology For Legal Process*, Chaired by D.R. Young at University of Denver in November 1970; see *Denver Law Journal*, 47, No. 4, 1970.

Appendix 1

92D CONGRESS
1ST SESSION

S. 2302

A BILL

To establish an Office of Technology Assessment for the Congress as an aid in the identification and consideration of existing and
probable impacts of technological application; to amend the National Science Foun-
daition Act of 1950; and for other purposes.

By Mr. JORDAN of North Carolina, Mr. ALLOTT,
Mr. KENNEDY, Mr. PASTORE, and Mr. PROUTY

JULY 19, 1971
Read twice and referred to the Committee on Rules
and Administration

S. 2302

IN THE SENATE OF THE UNITED STATES

JULY 19, 1971

Mr. JORDAN of North Carolina (for himself, Mr. ALLOTT, Mr. KENNEDY, Mr. PASTORE, and Mr. PROUTY) introduced the following bill; which was read twice and referred to the Committee on Rules and Administration

---

A BILL

To establish an Office of Technology Assessment for the Congress as an aid in the identification and consideration of existing and probable impacts of technological application; to amend the National Science Foundation Act of 1950; and for other purposes.

1   *Be it enacted by the Senate and House of Representa-*

2   *tives of the United States of America in Congress assembled,*

3   That this Act may be cited as the "Technology Assessment

4   Act of 1971".

5   DECLARATION OF PURPOSE

6   SEC. 2. The Congress hereby finds and declares that:

7   (a) Emergent national problems, physical, biological,

II

59

1 and social, are of such a nature and are developing at such

2 an unprecedented rate as to constitute a major threat to the

3 security and general welfare of the United States.

4   (b) Such problems are largely the result of and are

5 allied to—

6   (1) the increasing pressures of population;

7   (2) the rapid consumption of natural resources;

8   and

9   (3) the deterioration of the human environment,

10   natural and social,

11 though not necessarily limited to or by these factors.

12   (c) The growth in scale and extent of technological

13 application is a crucial element in such problems and either

14 is or can be a pivotal influence with respect both to their

15 cause and to their solution.

16   (d) The present mechanisms of the Congress do not

17 provide the legislative branch with adequate independent

18 and timely information concerning the potential applica-

19 tion or impact of such technology, particularly in those in-

20 stances where the Federal Government may be called upon

21 to consider support, management, or regulation of tech-

22 nological applications.

23   (e) It is therefore, imperative that the Congress equip

24 itself with new and effective means for securing competent,

25 unbiased information concerning the effects, physical, eco-

1 nomic, social, and political, of the applications of technology,

2 and that such information be utilized whenever appropriate

3 as one element in the legislative assessment of matters pend-

4 ing before the Congress.

5 ESTABLISHMENT OF THE OFFICE OF TECHNOLOGY

6 ASSESSMENT

7 SEC. 3. (a) In accordance with the rationale enunciated

8 in section 2, there is hereby created the Office of Technology

9 Assessment (hereinafter referred to as the "Office") which

10 shall be within and responsible to the legislative branch of

11 the Government.

12 (b) The Office shall consist of a Technology Assessment

13 Board (hereinafter referred to as the "Board") which shall

14 formulate and promulgate the policies of the Office, and a

15 Director who shall carry out such policies and administer

16 the operations of the Office.

17 (c) The basic responsibilities and duties of the Office

18 shall be to provide an early warning of the probable im-

19 pacts, positive and negative, of the applications of technology

20 and to develop other coordinate information which may

21 assist the Congress in determining the relative priorities of

22 programs before it. In carrying out such function, the Office

23 shall—

24 (1) identify existing or probable impacts of tech-

25 nology or technological programs;

1   (2) where possible establish cause-and-effect rela-
2 tionships;

3   (3) determine alternative technological methods of
4 implementing specific programs;

5   (4) determine alternative programs for achieving
6 requisite goals;

7   (5) make estimates and comparisons of the impacts
8 of alternative methods and programs;

9   (6) present findings of completed analyses to the
10 appropriate legislative authorities;

11   (7) identify areas where additional research or data
12 collection is required to provide adequate support for the
13 assessments and estimates described in paragraphs (1)
14 through (5) ; and

15   (8) undertake such additional associated tasks as
16 the appropriate authorities specified under subsection
17 (d) may direct.

18 (d) Activities undertaken by the Office may be initi-
19 ated by—

20   (1) the chairman of any standing, special, select,
21 or joint committee of the Congress;

22   (2) the Board; or

23   (3) the Director.

24 (e) Information, surveys, studies, reports, and findings
25 produced by the Office shall be made freely available to the

1 public except where (1) to do so would violate security

2 statutes, or (2) the information or other matter involved

3 could be withheld from the public, notwithstanding subsec-

4 tion (a) of section 552 of title 5, United States Code, under

5 one or more of the numbered paragraphs in subsection (b)

6 of such section.

7   (f) In undertaking the duties set out in subsection (c),

8 full use shall be made of competent personnel and organiza-

9 tions outside the Office, public or private; and special ad hoc

10 task forces or other arrangements may be formed by the

11 Director when appropriate.

12   TECHNOLOGY ASSESSMENT BOARD

13   SEC. 4. (a) The Board shall consist of eleven mem-

14 bers as follows:

15   (1) two Members of the Senate who shall not be

16   members of the same political party, to be appointed

17   by the President pro tempore of the Senate;

18   (2) two Members of the House of Representatives

19   who shall not be members of the same political party,

20   to be appointed by the Speaker of the House of Repre-

21   sentatives;

22   (3) the Comptroller General of the United States;

23   (4) the Director of the Congressional Research

24   Service of the Library of Congress;

25   (5) four members from the public, appointed by

1　　the President, by and with the advice and consent or

2　　the Senate, who shall be persons eminent in one or

3　　more fields of science or engineering or experienced in

4　　the administration of technological activities, or who may

5　　be judged qualified on the basis of contributions made

6　　to educational or public activities; and

7　　　　(6) the Director (except that he shall not be con-

8　　sidered a voting member for purposes of appointment or

9　　removal under the first sentence of section 5 (a)).

10　　(b) The Board, by majority vote, shall elect from

11　among its members appointed under subsection (a) (5) a

12　Chairman and a Vice Chairman, who shall serve for such

13　time and under such conditions as the Board may prescribe,

14　but for a period of not to exceed four years. In the absence

15　of the Chairman, or in the event of his incapacity, the Vice

16　Chairman shall fulfill the duties and functions of the

17　Chairman.

18　　(c) The Board shall meet upon the call of the Chair-

19　man or upon the petition of five or more of its members,

20　but it shall meet not less than twice each year.

21　　(d) Six members of the Board shall constitute a

22　quorum.

23　　(e) Any vacancy in the Board shall not affect its

24　powers, but shall be filled in the manner in which the vacant

25　position was originally filled.

64

1   (f) The term of office of each member of the Board
2 appointed under subsection (a) (5) shall be four years, ex-
3 cept that (1) any such member appointed to fill a vacancy
4 occurring prior to the expiration of the term for which his
5 predecessor was appointed shall be appointed for the re-
6 mainder of such term; and (2) the terms of office of such
7 members first taking office after the enactment of this Act
8 shall expire, as designated by the President at the time of
9 appointment, two at the end of two years and two at the
10 end of four years, after the date of the enactment of this
11 Act. No person shall be appointed a member of the Board
12 under subsection (a) (5) more than twice.

13   (g) (1) The members of the Board other than those
14 appointed under subsection (a) (5) shall receive no compen-
15 sation for their services as members of the Board, but shall be
16 allowed necessary travel expenses (or, in the alternative,
17 mileage for use of privately owned vehicles and a per diem
18 in lieu of subsistence not to exceed the rates prescribed in
19 sections 5702 and 5704 of title 5, United States Code), and
20 other necessary expenses incurred by them in the perform-
21 ance of duties vested in the Board, without regard to the
22 provisions of subchapter I of chapter 57 of title 5, United
23 States Code, the Standardized Government Travel Regula-
24 tions, or section 5731 of title 5, United States Code.

25   (2) The members of the Board appointed under sub-

1   section (a) (5) shall each receive compensation at the rate

2   of $100 for each day engaged in the actual performance of

3   duties vested in the Board, and in addition shall be reim-

4   bursed for travel, subsistence, and other necessary expenses

5   in the manner provided in paragraph (1) of this subsection.

6   DIRECTOR AND DEPUTY DIRECTOR

7   SEC. 5. (a) The Director of the Office of Technology

8   Assessment shall be appointed by the Board and shall serve

9   for a term of six years unless sooner removed by the Board.

10   He shall receive basic pay at the rate provided for level II of

11   the Executive Schedule under section 5313 of title 5, United

12   States Code.

13   (b) In addition to the powers and duties vested in him

14   by this Act, the Director shall exercise such powers and

15   duties as may be delegated to him by the Board.

16   (c) The Director may appoint, with the approval of the

17   Board, a Deputy Director who shall perform such functions

18   as the Director may prescribe and who shall be Acting Direc-

19   tor during the absence or incapacity of the Director or in the

20   event of a vacancy in the office of Director. The Deputy

21   Director shall receive basic pay at the rate provided for

22   level III of the Executive Schedule under section 5314 of

23   title 5, United States Code.

24   (d) Neither the Director nor the Deputy Director shall

25   engage in any other business, vocation, or employment than

1 that of serving as such Director or Deputy Director, as the
2 case may be; nor shall the Director or Deputy Director, ex-
3 cept with the approval of the Board, hold any office in, or
4 act in any capacity for, any organization, agency, or institu-
5 tion with which the Office makes any contract or other
6 arrangement under this Act.

7 AUTHORITY OF THE OFFICE

8 SEC. 6. (a) The Office shall have the authority, within
9 the limits of available appropriations, to do all things neces-
10 sary to carry out the provisions of this Act, including, but
11 without being limited to, the authority to—

12   (1) prescribe such rules and regulations as it deems
13   necessary governing the manner of its operation and its
14   organization and personnel;

15   (2) make such expenditures as may be necessary
16   for administering the provisions of this Act;

17   (3) enter into contracts or other arrangements as
18   may be necessary for the conduct of its work with any
19   agency or instrumentality of the United States, with any
20   foreign country or international agency, with any State,
21   territory, or possession or any political subdivision
22   thereof, or with any person, firm, association, corpora-
23   tion, or educational institution, with or without reim-
24   bursement, without performance or other bonds, and

S. 2302——2

1   without regard to section 3709 of the Revised Statutes

2   (41 U.S.C. 5) ;

3   (4) make advance, progress, and other payments

4   which relate to technology assessment without regard

5   to the provisions of section 3648 of the Revised Statutes

6   (31 U.S.C. 529) ;

7   (5) acquire by purchase, lease, loan, or gift, and

8   holds and dispose of by sale, lease, or loan, real and per-

9   sonal property of all kinds necessary for, or resulting

10   from, the exercise of authority granted by this Act; and

11   (6) accept and utilize the services of voluntary and

12   uncompensated personnel and provide transportation and

13   subsistence as authorized by section 5703 of title 5,

14   United States Code, for persons serving without

15   compensation.

16   (b) The Director shall, in accordance with such policies

17 as the Board shall prescribe, appoint and fix the compensa-

18 tions of such personnel as may be necessary to carry out the

19 provisions of this Act. Such appointments shall be made and

20 such compensation shall be fixed in accordance with the pro-

21 visions of title 5, United States Code, governing appoint-

22 ments in the competitive service, and the provisions of chap-

23 ter 51 and subchapter III of chapter 53 of such title relating

24 to classification and General Schedule pay rates; except that

25 the Director may, in accordance with such policies as the

1 Board shall prescribe, employ such technical and professional

2 personnel and fix their compensation without regard to such

3 provisions as he may deem necessary for the discharge of the

4 responsibilities of the Office under this Act.

5 (c) The Office shall not, itself, operate any laboratories,

6 pilot plants, or test facilities in the pursuit of its mission.

7 (d) (1) The Office or (on the authorization of the Of-

8 fice) any of its duly constituted officers may, for the purpose

9 of carrying out the provisions of this Act, hold such hearings,

10 take such testimony, and sit and act at such times and places

11 as the Office deems advisable. For this purpose the Office is

12 authorized to require the attendance of such persons and the

13 production of such books, records, documents, or data, by

14 subpena or otherwise, and to take such testimony and rec-

15 ords, as it deems necessary. Subpenas may be issued by the

16 Director or by any person designated by him. If compliance

17 with such a subpena by the person to whom it is issued or

18 upon whom it is served would (in such person's judgment)

19 require the disclosure of trade secrets or other commercial,

20 financial, or proprietary information which is privileged or

21 confidential, or constitute a clearly unwarranted invasion of

22 privacy, such person may petition the United States district

23 court for the district in which he resides or has his principal

24 place of business, or in which the books, records, documents,

25 or data involved are situated, and such court (after inspect-

1 ing such books, records, documents, or data in camera) may

2 excise and release from the subpena any portion thereof

3 which it determines would require such disclosure or con-

4 stitute such invasion. Where the subpena or such portion

5 thereof would require such disclosure or constitute such in-

6 vasion but the books, records, documents, or data involved are

7 shown to be germane to the matters under consideration and

8 necessary for the effective conduct by the Office of its pro-

9 ceedings or deliberations with respect thereto, the court may

10 require that such books, records, documents, or data be

11 produced or made available to the Office in accordance with

12 the subpena but subject to such conditions and limitations

13 of access as will prevent their public disclosure and protect

14 their confidentiality.

15   (2) In case of contumacy or disobedience to a subpena

16 issued under paragraph (1) the Attorney General, at the

17 request of the Office, shall invoke the aid of the United States

18 district court for the district in which the person to whom

19 the subpena was issued or upon whom it was served resides

20 or has his principal place of business, or in which the books,

21 records, documents, or data involved are situated, or the aid

22 of any other United States district court within the jurisdic-

23 tion of which the Office's proceedings are being carried on,

24 in requiring the production of such books, records, documents,

25 or data or the attendance and testimony of such person in

1 accordance with the subpena (subject to any conditions or

2 limitations of access which may have been imposed by such

3 court or any other court under the last sentence of para-

4 graph (1)). Such court may issue an order requiring the

5 person to whom the subpena was issued or upon whom it was

6 served to produce the books, records, documents, or data

7 involved, or to appear and testify, or both, in accordance

8 with the subpena (subject to any such conditions or limita-

9 tions of access) ; and any failure to obey such order of the

10 court may be punished by the court as a contempt thereof.

11 (e) Each department, agency, or instrumentality of

12 the executive branch of the Government, including inde-

13 pendent agencies, is authorized and directed to furnish to the

14 Office, upon request by the Director, such information as

15 the Office deems necessary to carry out its functions under

16 this Act.

17 (f) Contractors and other parties entering into contracts

18 and other arrangements under this section which involve

19 cost to the Government shall maintain such books and

20 related records as will facilitate an effective audit in such

21 detail and in such manner as shall be prescribed by the

22 Director, and such books and records (and related documents

23 and papers) shall be available to the Director and the

24 Comptroller General or any of their duly authorized repre-

25 sentatives for the purpose of audit and examination.

UTILIZATION OF THE LIBRARY OF CONGRESS

SEC. 7. (a) Pursuant to the objectives of this Act, the Librarian of Congress is authorized to make available to the Office such services and assistance by the Congressional Research Service as may be appropriate and feasible.

(b) The foregoing services and assistance to the Office shall include all of the services and assistance which the Congressional Research Service is presently authorized to provide to the Congress, and shall particularly include, without being limited to, the following:

(1) maintaining a monitoring indicator system with respect to the natural and social environments which might reveal early impacts of technological change, but any such system shall be coordinated with other assessment activities which may exist in the departments and agencies of the executive branch of the Government;

(2) making surveys of ongoing and proposed programs of government with a high or novel technology content, together with timetables of applied science showing promising developments;

(3) publishing, from time to time, anticipatory reports and forecasts;

1   (4) recording the activities and responsibilities of

2   Federal agencies in affecting or being affected by tech-

3   nological change;

4   (5) when warranted, recommending full-scale as-

5   sessments;

6   (6) preparing background reports to aid in receiv-

7   ing and using the assessments;

8   (7) providing staff assistance in preparing for or

9   * holding committee hearings to consider the findings of

10   the assessments;

11   (8) reviewing the findings of any assessment made

12   by or for the Office; and

13   (9) assisting the Office in the maintenance of liai-

14   son with executive agencies involved in technology

15   assessments.

16   (c) Nothing in this section shall alter or modify any

17 services or responsibilities other than those performed for

18 the Office, which the Congressional Research Service under

19 law performs for or on behalf of the Congress. The Librarian

20 is, however, authorized to establish within the Congressional

21 Research Service such additional divisions, groups, or other

22 organizational entities as may be necessary to carry out

23 the objectives of this Act, including the functions enumer-

24 ated in this section.

1   (d) Services and assistance made available to the Office

2 by the Congressional Research Service in accordance with

3 this section may be provided with or without reimbursement

4 from funds of the Office, as agreed upon by the Chairman of

5 the Board and the Librarian of Congress.

6 COORDINATION WITH THE NATIONAL SCIENCE FOUNDATION

7   SEC. 8. (a) The Office shall maintain a continuing liaison

8 with the National Science Foundation with respect to—

9   (1) grants and contracts formulated or activated

10   by the Foundation which are for purposes of technology

11   assessment, and

12   (2) the promotion of coordination in areas of tech-

13   nology assessment, and the avoidance of unnecessary

14   duplication or overlapping of research activities in the

15   development of technology assessment techniques and

16   programs.

17   (b) Section 3 (b) of the National Science Foundation

18 Act of 1950, as amended, is hereby amended to read as

19 follows:

20   "(b) The Foundation is authorized to initiate and sup-

21 port specific scientific activities in connection with matters

22 relating to international cooperation, national security, and

23 the effects of scientific applications upon society by making

24 contracts or other arrangements (including grants, loans, and

25 other forms of assistance) for the conduct of such activities.

74

1 When initiated or supported pursuant to requests made by

2 any other Federal department or agency, including the Office

3 of Technology Assessment, such activities shall be financed

4 whenever feasible from funds transferred to the Foundation

5 by the requesting official as provided in section 14 (g), and

6 any such activities shall be unclassified and shall be identi-

7 fied by the Foundation as being undertaken at the request

8 of the appropriate official."

9 ANNUAL REPORT

10   SEC. 9. The Office shall submit to the Congress and to

11 the President an annual report which shall, among other

12 things, evaluate the existing state of the art with regard to

13 technology assessment techniques and forecast, insofar as

14 may be feasible, technological areas requiring future atten-

15 tion. The report shall be submitted not later than March 15

16 each year.

17 FINANCIAL AND ADMINISTRATIVE SERVICES

18   SEC. 10. Financial and administrative services (includ-

19 ing those related to budgeting, accounting, financial report-

20 ing, personnel, and procurement) shall be provided the

21 Office by the General Accounting Office, with or without

22 reimbursement from funds of the Office, as may be agreed

23 upon by the Chairman of the Board and the Comptroller

24 General of the United States. The regulations of the General

25 Accounting Office for the collection of indebtedness of person-

1 nel resulting from erroneous payments (under section 5514

2 (b) of title 5, United States Code) shall apply to the col-

3 lection of erroneous payments made to or on behalf of an

4 Office employee, and the regulations of the Comptroller

5 General for the administrative control of funds (under sec-

6 tion 3679 (g) of the Revised Statutes (31 U.S.C. 665 (g))

7 shall apply to appropriations of the Office; and the Office

8 shall not be required to prescribe such regulations.

9 APPROPRIATIONS

10 SEC. 11. (a) To enable the Office to carry out its

11 powers and duties, there is hereby authorized to be appro-

12 priated to the Office, out of any money in the Treasury not

13 otherwise appropriated, not to exceed $5,000,000 for the

14 fiscal year ending June 30, 1972, and thereafter such sums

15 as may be necessary.

16 (b) Appropriations made pursuant to the authority pro-

17 vided in subsection (a) shall remain available for obligation,

18 for expenditure, or for obligation and expenditure for such

19 period or periods as may be specified in the Act making such

20 appropriations.

Appendix 2

92D CONGRESS
2D SESSION # H. R. 10243

AN ACT

To establish an Office of Technology Assessment for the Congress as an aid in the identification and consideration of existing and probable impacts of technological application; to amend the National Science Foundation Act of 1950; and for other purposes.

FEBRUARY 9, 1972

Read twice and referred to the Committee on Rules and Administration

92D CONGRESS
2D SESSION

H. R. 10243

IN THE SENATE OF THE UNITED STATES

FEBRUARY 9, 1972

Read twice and referred to the Committee on Rules and Administration

AN ACT

To establish an Office of Technology Assessment for the Congress
as an aid in the identification and consideration of existing
and probable impacts of technological application; to amend
the National Science Foundation Act of 1950; and for other
purposes.

1 *Be it enacted by the Senate and House of Representa-*

2 *tives of the United States of America in Congress assembled,*

3 That this Act may be cited as the "Technology Assessment

4 Act of 1972".

5   DECLARATION OF PURPOSE

6 SEC. 2. The Congress hereby finds and declares that:

7 (a) Emergent national problems, physical, biological,

8 and social, are of such a nature and are developing at such

VI—O

1 an unprecedented rate as to constitute a major threat to the
2 security and general welfare of the United States.

3   (b) Such problems are largely the result of and are
4 allied to—

5   (1) the increasing pressures of population;

6   (2) the rapid consumption of natural resources;
7   and

8   (3) the deterioration of the human environment,
9   natural and social,

10 though not necessarily limited to or by these factors.

11   (c) The growth in scale and extent of technological
12 application is a crucial element in such problems and either
13 is or can be a pivotal influence with respect both to their
14 cause and to their solution.

15   (d) The present mechanisms of the Congress do not
16 provide the legislative branch with adequate independent
17 and timely information concerning the potential application
18 or impact of such technology, particularly in those instances
19 where the Federal Government may be called upon to
20 consider support, management, or regulation of technological
21 applications.

22   (e) It is therefore imperative that the Congress equip
23 itself with new and effective means for securing competent,
24 unbiased information concerning the effects, physical, eco-
25 nomic, social, and political, of the applications of technology,

1 and that such information be utilized whenever appropriate

2 as one element in the legislative assessment of matters

3 pending before the Congress.

4 ESTABLISHMENT OF THE OFFICE OF TECHNOLOGY

5 ASSESSMENT

6 SEC. 3. (a) In accordance with the rationale enunciated

7 in section 2, there is hereby created the Office of Technology

8 Assessment (hereinafter referred to as the "Office") which

9 shall be within and responsible to the legislative branch of the

10 Government.

11 (b) The Office shall consist of a Technology Assessment

12 Board (hereinafter referred to as the "Board") which shall

13 formulate and promulgate the policies of the Office, and a

14 Director who shall carry out such policies and administer

15 the operations of the Office.

16 (c) The basic responsibilities and duties of the Office

17 shall be to provide an early warning of the probable im-

18 pacts, positive and negative, of the applications of technology

19 and to develop other coordinate information which may

20 assist the Congress in determining the relative priorities of

21 programs before it. In carrying out such function, the Office

22 shall—

23 (1) identify existing or probable impacts of tech-

24 nology or technological programs;

1   (2) where possible establish cause-and-effect rela-
2 tionships;

3   (3) determine alternative technological methods of
4 implementing specific programs;

5   (4) determine alternative programs for achieving
6 requisite goals;

7   (5) make estimates and comparisons of the impacts
8 of alternative methods and programs;

9   (6) present findings of completed analyses to the
10 appropriate legislative authorities;

11   (7) identify areas where additional research or data
12 collection is required to provide adequate support for the
13 assessments and estimates described in paragraphs (1)
14 through (5) ; and

15   (8) undertake such additional associated tasks as
16 the appropriate authorities specified under subsection
17 (d) may direct.

18   (d) Activities undertaken by the Office may be initi-
19 ated by—

20   (1) the chairman of any standing, special, select,
21 or joint committee of the Congress, acting for himself
22 or at the request of the ranking minority member or a
23 majority of the committee members; or

24   (2) the Board.

25   (e) Information, surveys, studies, reports, and findings

1 produced by the Office shall be made freely available to the

2 public except where (1) to do so would violate security

3 statutes, or (2) the information or other matter involved

4 could be withheld from the public, notwithstanding subsec-

5 tion (a) of section 552 of title 5, United States Code, under

6 one or more of the numbered paragraphs in subsection (b)

7 of such section.

8   (f) In undertaking the duties set out in subsection (c),

9 full use shall be made of competent personnel and organiza-

10 tions outside the Office, public or private; and special ad hoc

11 task forces or other arrangements may be formed by the

12 Director when appropriate.

13   TECHNOLOGY ASSESSMENT BOARD

14   SEC. 4. (a) The Board shall consist of ten members as

15 follows:

16   (1) five Members of the Senate, appointed by the

17   President pro tempore of the Senate, three from the

18   majority party and two from the minority party; and

19   (2) five Members of the House of Representatives

20   appointed by the Speaker of the House of Representa-

21   tives, three from the majority party and two from the

22   minority party.

23   (c) Vacancies in the membership of the Board shall not

24 affect the power of the remaining members to execute the

1 functions of the Board and shall be filled in the same manner

2 as in the case of the original appointment.

3   (d) The Board shall select a chairman and a vice chair-

4 man from among its members at the beginning of each Con-

5 gress. The vice chairman shall act in the place and stead of

6 the chairman in the absence of the chairman. The chairman-

7 ship and the vice chairmanship shall alternate between the

8 Senate and the House of Representatives with each Congress.

9 The chairman during each even-numbered Congress shall be

10 selected by the Members of the House of Representatives on

11 the Board from among their number. The vice chairman

12 during each Congress shall be chosen in the same manner

13 from that House of Congress other than the House of Con-

14 gress of which the chairman is a Member.

15   DIRECTOR AND DEPUTY DIRECTOR

16   SEC. 5. (a) The Director of the Office of Technology

17 Assessment shall be appointed by the Board and shall serve

18 for a term of six years unless sooner removed by the Board.

19 He shall receive basic pay at the rate provided for level II

20 of the Executive Schedule under section 5313 of title 5,

21 United States Code.

22   (b) In addition to the powers and duties vested in him

23 by this Act, the Director shall exercise such powers and

24 duties as may be delegated to him by the Board.

25   (c) The Director may appoint, with the approval of the

1 Board, a Deputy Director who shall perform such functions
2 as the Director may prescribe and who shall be Acting Di-
3 rector during the absence or incapacity of the Director or in
4 the event of a vacancy in the office of Director. The Deputy
5 Director shall receive basic pay at the rate provided for
6 level III of the Executive Schedule under section 5314 of
7 title 5, United States Code.

8 (d) Neither the Director nor the Deputy Director shall
9 engage in any other business, vocation, or employment than
10 that of serving as such Director or Deputy Director, as the
11 case may be; nor shall the Director or Deputy Director,
12 except with the approval of the Board, hold any office in,
13 or act in any capacity for, any organization, agency, or
14 institution with which the Office makes any contract or
15 other arrangement under this Act.

16 AUTHORITY OF THE OFFICE

17 SEC. 6. (a) The Office shall have the authority, within
18 the limits of available appropriations, to do all things neces-
19 sary to carry out the provisions of this Act, including, but
20 without being limited to, the authority to—

21 (1) prescribe such rules and regulations as it deems
22 necessary governing the manner of its operation and its
23 organization and personnel;

24 (2) make such expenditures as may be necessary
25 for administering the provisions of this Act;

1   (3) enter into contracts or other arrangements as

2 may be necessary for the conduct of its work with any

3 agency or instrumentality of the United States, with any

4 foreign country or international agency, with any State,

5 territory, or possession or any political subdivision there-

6 of, or with any person, firm, association, corporation, or

7 educational institution, with or without reimbursement,

8 without performance or other bonds, and without regard

9 to section 3709 of the Revised Statutes (41 U.S.C. 5) ;

10   (4) make advance, progress, and other payments

11 which relate to technology assessment without regard

12 to the provisions of section 3648 of the Revised Statutes

13 (31 U.S.C. 529) ;

14   (5) acquire by purchase, lease, loan, or gift, and

15 hold and dispose of by sale, lease, or loan, real and

16 personal property of all kinds necessary for or resulting

17 from, the exercise of authority granted by this Act; and

18   (6) accept and utilize the services of voluntary and

19 uncompensated personnel and provide transportation and

20 subsistence as authorized by section 5703 of title 5,

21 United States Code, for persons serving without com-

22 pensation.

23 (b) The Director shall, in accordance with such policies

24 as the Board shall prescribe, appoint and fix the compensa-

1 tion of such personnel as may be necessary to carry out the
2 provisions of this Act. Such appointments shall be made and
3 such compensation shall be fixed in accordance with the pro-
4 visions of title 5, United States Code, governing appoint-
5 ments in the competitive service, and the provisions of chap-
6 ter 51 and subchapter III of chapter 53 of such title relating
7 to classification and General Schedule pay rates.

8   (c) The Office shall not, itself, operate any laboratories,
9 pilot plants, or test facilities in the pursuit of its mission.

10   (e) Each department, agency, or instrumentality of
11 the executive branch of the Government, including inde-
12 pendent agencies, is authorized and directed to furnish to
13 the Office, upon request by the Director, such information
14 as the Office deems necessary to carry out its functions under
15 this Act.

16   (f) Contractors and other parties entering into contracts
17 and other arrangements under this section which involve
18 costs to the Government shall maintain such books and re-
19 lated records as will facilitate an effective audit in such detail
20 and in such manner shall be prescribed by the Director, and
21 such books and records (and related documents and papers)
22 shall be available to the Director and the Comptroller General
23 or any of their duly authorized representatives for the pur-
24 pose of audit and examination.

1 UTILIZATION OF THE LIBRARY OF CONGRESS

2 SEC. 7. (a) Pursuant to the objectives of this Act, the

3 Librarian of Congress is authorized to make available to the

4 Office such services and assistance by the Congressional Re-

5 search Service as may be appropriate and feasible.

6 (b) The foregoing services and assistance to the Office

7 shall include all of the services and assistance which the

8 Congressional Research Service is presently authorized to

9 provide to the Congress, and shall particularly include, with-

10 out being limited to, the following:

11 (1) maintaining a monitoring indicator system with

12 respect to the natural and social environments which

13 might reveal early impacts of technological change, but

14 any such system shall be coordinated with other assess-

15 ment activities which may exist in the departments and

16 agencies of the executive branch of the Government;

17 (2) making surveys of ongoing and proposed pro-

18 grams of government with a high or novel technology

19 content, together with timetables of applied science

20 showing promising developments;

21 (3) publishing, from time to time, anticipatory

22 reports and forecasts;

23 (4) recording the activities and responsibilities of

24 Federal agencies in affecting or being affected by tech-

25 nological change;

1   (5) when warranted, recommending full-scale as-

2   sessments;

3   (6) preparing background reports to aid in re-

4   ceiving and using the assessments;

5   (7) providing staff assistance in preparing for or

6   holding committee hearings to consider the findings of

7   the assessments;

8   (8) reviewing the findings of any assessment made

9   by or for the Office; and

10   (9) assisting the Office in the maintenance of liaison

11   with executive agencies involved in technology assess-

12   ments.

13   (c) Nothing in this section shall alter or modify any

14 services or responsibilities, other than those performed for

15 the Office, which the Congressional Research Service under

16 law performs for or on behalf of the Congress. The Librarian

17 is, however, authorized to establish within the Congressional

18 Research Service such additional divisions, groups, or other

19 organization entities as may be necessary to carry out the

20 objectives of this Act, including the functions enumerated in

21 this section.

22   (d) Services and assistance made available to the Office

23 by the Congressional Research Service in accordance with

24 this section may be provided with or without reimbursement

1   from funds of the Office, as agreed upon by the Chairman

2   of the Board and the Librarian of Congress.

3   COORDINATION WITH THE NATIONAL SCIENCE

4   FOUNDATION

5   SEC. 8. (a) The Office shall maintain a continuing liaison

6   with the National Science Foundation with respect to—

7   (1) grants and contracts formulated or activated

8   by the Foundation which are for purposes of technology

9   assessment, and

10   (2) the promotion of coordination in areas of tech-

11   nology assessment, and the avoidance of unnecessary

12   duplication or overlapping of research activities in the

13   development of technology assessment techniques and

14   programs.

15   (b) Section 3 (b) of the National Science Foundation

16   Act of 1950, as amended, is hereby amended to read as

17   follows:

18   "(b) The Foundation is authorized to initiate and sup-

19   port specific scientific activities in connection with matters

20   relating to international cooperation, national security, and

21   the effects of scientific applications upon society by making

22   contracts or other arrangements (including grants, loans, and

23   other forms of assistance) for the conduct of such activities.

24   When initiated or supported pursuant to requests made by

25   any other Federal department or agency, including the

1 Office of Technology Assessment, such activities shall be
2 financed whenever feasible from funds transferred to the
3 Foundation by the requesting official as provided in section
4 14 (g), and any such activities shall be unclassified and shall
5 be identified by the Foundation as being undertaken at the
6 request of the appropriate official."

7 ANNUAL REPORT

8 SEC. 9. The Office shall submit to the Congress and to
9 the President an annual report which shall, among other
10 things, evaluate the existing state of the art with regard to
11 technology assessment techniques and forecast, insofar as
12 may be feasible, technological areas requiring future atten-
13 tion. The report shall be submitted not later than March 15
14 each year.

15 UTILIZATION OF THE GENERAL ACCOUNTING OFFICE

16 SEC. 10. Financial and administrative services (includ-
17 ing those related to budgeting, accounting, financial report-
18 ing, personnel, and procurement) and such other services
19 as may be appropriate shall be provided the Office by the
20 General Accounting Office, with or without reimbursement
21 from funds of the Office, as may be agreed upon by the
22 Chairman of the Board and the Comptroller General of the
23 United States. The regulations of the General Accounting
24 Office for the collection of indebtedness of personnel resulting
25 from erroneous payments (under section 5514 (b) of title 5,

1 United States Code) shall apply to the collection of erro-
2 neous payments made to or on behalf of an Office employee,
3 and the regulations of the Comptroller General for the ad-
4 ministrative control of funds (under section 3679 (g)) of
5 the Revised Statutes (31 U.S.C. 665 (g)) shall apply to
6 appropriations of the Office; and the Office shall not be
7 required to prescribe such regulations.

8   APPROPRIATIONS

9   SEC. 11. (a) To enable the Office to carry out its
10 powers and duties, there is hereby authorized to be appro-
11 priated to the Office, out of any money in the Treasury not
12 otherwise appropriated, not to exceed $5,000,000 in the
13 aggregate for the two fiscal years ending June 30, 1973,
14 and June 30, 1974.

15   (b) Appropriations made pursuant to the authority pro-
16 vided in subsection (a) shall remain available for obligation,
17 for expenditure, or for obligation and expenditure for such
18 period or periods as may be specified in the Act making such
19 appropriations.

Passed the House of Representatives February 8, 1972.

Attest: W. PAT JENNINGS,
                                    Clerk.

CHAPTER 3

Technology Assessment in the U.K. Public Service

"My conclusion is that a great work of persuasion will be needed, directed equally at politicians and at corporation presidents before technology autonomous can be defeated and before the great pool of creative technologists can direct their inventiveness to the problems which really matter."

Dennis Gabor (1972)

Introduction

Viewed from without, the decision-making procedure in U.K. government and public agencies appear labyrinthine, particularly when the problems under review have any technological content. Attempts at elucidating the process involved by means of interview or questionnaire are usually frustrated because of the semantic difficulties of defining "technology assessment" and its associated techniques[1].

---

[1] It was my own attempt, through interview and questionnaire, to obtain cohesive information from government agencies that convinced me that a proper discourse could not be achieved without the use of a comprehensive "primer" on technology assessment. The Civil Service College, Sunningdale, does teach subjects which are part of assessment technique but apparently the College's good services have not been operating long enough to have familiarized many civil servants with nomenclature and procedures used in technology assessment.

From my own research and knowledge of the public service, it has only been possible to derive a subjectively biased collection of relevant information, notwithstanding the respondent agencies being understandably cautious because of the debate raging about the report of Lord Rothschild's Central Policy Review Staff on "A Framework for Government Research and Development", Cmnd 4814. Without revealing information given in confidence, this chapter will contain some indication of present competence in, and attitude towards, technology assessment in the public service. In what follows, some expurgation of real sources is necessary and "blanket" attribution must be used. Because the assessment activity of the Programmes Analysis Unit is, within the customary time for Civil Service change, an innovation, it forms the subject of the next chapter.

Does the Service Ever Explicitly Consider Technology Impact?

Many mandibles are available to the public service for chewing over problems and therefore the answer to the general question heading this section must be "yes". On the other hand, answers to the question (and all the others nested within it) are not easy to obtain from individual departments. A properly conducted survey of decision-making procedures and their analytic inputs is quite obviously an horrendous task and would require an inquisition of Royal Commissions supported by a priesthood of unimpeachable analysts. Some answers to general questions are, nevertheless, possible (on a personal basis) and the following very loose statistics may give readers some perception of what the present position, very approximately, is.

54% of a subjectively chosen, but confidential, sample of public departments and government agencies claim to use a formalized procedure for the evaluation of impacts of new technology,

23% claim that they have no formal procedure, and

23% do not wish to answer the question.

38% of organizations operating formalized procedures are concerned with both economic and social impact,

16% are concerned exclusively with economic impact,

8% are concerned exclusively with social impact, and

38% do not define the impact they are concerned with.

8% of the organizations operating formalized procedures have permanent teams working upon the evaluation of possible impacts,

8% use ad hoc teams when the need to evaluate impact arises,
38% use both permanent and ad hoc teams, and
48% do not define what kind of teams they use.

76% have an informal procedure for the evaluation of technological im-
     pacts,
16% have not got an informal procedure, and
8% do not discuss procedure.

77% of teams carrying out evaluations of impact have multi-disciplinary
     structures,
8% claim to have teams of single discipline, and
15% do not specify whether they are multi-disciplinary or not.

43% make use of outside consultants for the evaluation of technological
     impact, and
57% do not use consultants.

approx. 100% of those employing consultants select them from within and
without the public service.

50% of potential benefits examined exceed £10 million,
25% lie between £1 million and £10 million,
12.5% lie between £100 thousand and £1 million, and
12.5% lie between £10 thousand and £100 thousand.

17% of the anticipated costs exceed £10 million,
33% lie between £1 million and £10 million,
33% lie between £100 thousand and £1 million, and
17% lie between £10 thousand and £100 thousand.

28% have an evaluation cost greater than £1 million,
14% lie between £100 thousand and £1 million,
14% lie between £10 thousand and £100 thousand, and
44% lie between £1 thousand and £10 thousand.

From the above table it is possible to obtain a loose indicator of
the "formalized" concern with economic and social benefits and
disbenefits of technology, as follows.

(a) Not less than 0.54 (38 + 8) = 25% of the sample have *formalized* procedures for the evaluation of *social* impacts of new
technology.

(b) Not less than 0.54 (38 + 16) = 29% of the sample have *formalized* procedures for the evaluation of *economic* impacts of new
technology.

Another "statistic"of relevance is that about (38 + 8) = 46% of
the sample have permanent teams working on formalized procedures and that these teams are about 77% multi-disciplinary.

94

Rather surprisingly, 43% use outside consultants and these are about equally split between consultants from within and without the public service.

Comments

It is certainly gratifying, to those who do not wish to measure all progress by economic yardsticks, to note the overall percentages "formally" working on economic and social impacts are, although small, approximately equal. Their smallness is mitigated by the respondent's claim that 76% of them have "informal" procedures for the evaluation of technological impacts. Nevertheless, some residual concern may be left by the realization that 50% of the respondents deal with potential benefits in excess of £10 million and 17% claim to examine potential costs also in excess of £10 million.

Information obtained about the techniques used during analysis of possible impacts shows (not surprisingly) nearly every possible, and known, technique to be in use. No useful purpose would be served by airing even the approximate distribution of the claimed use of these techniques because one man's operational research, or benefit—cost analysis, is another man's arithmetic or obscurantism until proven otherwise. An appraisal of the expertise with which the techniques have been used would keep an analysis stewing in other people's outpourings for a lifetime. What can be said is that it is indeed remarkable, if all the claimed techniques really are available to the public service, that there is such a dearth of demonstrations of use whenever possible social, or economic, impacts are discussed by the public media. We hear, for example, that supersonic transports could, through the effect of their condensation trails, cause a degeneration of stratospheric ozone, thus increasing the incidence on earth of potentially harmful ultraviolet radiation. Yet we are informed that the effect would be well within the daily natural variations and not be of much significance. However much I trust this assertion to be correct, because of access to scientific writings, it would be very reassuring to read even a literary account in the U.K. public press of how the best techniques of analysis substantiate this claim [1].

---

[1] I personally support Concorde at this stage in its development but I am appalled by the way the public has had the new SST technology forced upon it without being treated as if they were really capable of understanding the ramifications of supersonic flight.

It is possible that a new kind of technological hypochondria could be generated by serving the public with a distillation of conclusions derived by the use of evaluation techniques, but when the public ultimately pay the social costs of mis-applied technology, they have the right to become hypochondriacs if they so chose.

Naturally, a few of the respondents in my one-man investigation expressed scepticism about some of the newer and less proven techniques available to the assessor of technology. One respondent speaking from a senior position in a "technically based unit", claimed to be "extremely sceptical as to the benefits that can accrue from a professional approach to technological forecasting".

Another respondent made the ambiguous statement "it has taken a little while to reply to your letter about methods used in the Public Service for the evaluation of new technology because your enquiry does not really apply to our Organization whose business is research". This reply would have been partially understandable if the Organization had been working only on the frontiers of curiosity-motivated research but, startlingly, it came from an organization primarily concerned with applied research and specialized technology.

By comparison, some of the most non-research-like departments sometimes showed a commendable awareness of the usefulness of new technology in facilitating their day-to-day business. An organization concerned exclusively with social science responded that they were not directly involved in the evaluation of new technology but did support some work of this sort.

Organizations involved in defence work naturally emphasized their involvement with "continuous cost effectiveness comparisons" but one did report that "for the small amount of non-defence work the scientist concerned is encouraged to do his own studies" of economic and social impact "and if the advantage is not clear cut, we tend to withdraw or postpone". Prima facie this is a reasonable attitude because unless the originator of the work is prepared to claim a benefit, the project will probably not survive against outside criticism. Yet how many potentially useful projects from defence work have not seen the light-of-day because the in-house analyst was incompetent or too hard pressed to make a thorough evaluation?

The point made earlier about being handicapped in my independent research by the CPRS study is summed up in the following quotation by a Director of a major research establishment.

96

"The methods to be used are at present under consideration by the Government and I understand that there will shortly be a White Paper on the subject. In these circumstances I feel it would be inappropriate at the moment for me to deal with your questionnaire".

Semantic difficulties are well illustrated by the establishment, concerned with biology, which claimed not to be concerned with "technological R & D, but with biological investigations on which to base advice....."

Quite clearly this organization has recognized a dichotomy between technology and biology where none should exist. At a time when biology is becoming more and more of interest to the technologist and bio-engineering is looming large in the thoughts of the most perceptive thinkers, ivory-tower biologists (including those privileged in giving advice) would do well to remember that the generally accepted definition of technology (see Galbraith, 1967) is

"the systematic application of scientific and other organized knowledge to practical tasks".

Lord Rothschild's Emporium of Science

The growing groundswell of public concern about governmental handling of science and technology has created many ripples and some spindrift, with conversational spray reducing visibility. Prime Minister Heath's administration, with an eye on wider issues, had formulated ideas about a Central Policy Review Staff (CPRS) when in opposition and thought that a "think-tank" would help central government apply certain "ideas" from the management sciences to collective issues.

Shortly after the 1970 election, Lord Rothschild, head of the British Government's innovatory CPRS "think-tank", with the help of

"15 extremely able, rather analytical types who can look at any proposition independently, *given that it is not "linguistically too technical nor too specialized"*[1],

studied and published a contribution to discussion Green Paper

---

[1] See Angela Cromme's interview with Lord Rothschild, "On being a Grand Vizier", reported in the New Scientist, 27 April 1972 (the emphasis is mine).

(Cmnd 4814) on "A Framework for Governmental Research and Development". The CPRS's brave (or insouciant) presentation of their recommendations appeared to have been made without studying all the evidence, for unsettling the scientific community, and they were rebuked from some quarters (see Benn, 1972). At a time when American Congressmen were supporting Bill HR 10243 (see the previous chapter) to ameliorate the worst technological excesses of the market place, CPRS had, in fact, mounted what seemed to many to be an attack on a British system which, despite some fragmentation, has some of the desirable features of the proposed American Office of Technology assessment [1].

Apart from an unprecedented attack on the British research council system, the gist of the CPRS recommendation was that there should be a customer—contractor sytem (an internal market for technology) within the British public service for the control of expenditure on technological and applied scientific research. The CPRS suggestion is that Chief Scientists and their staffs be deputed to represent public customers for the services and products of applied research and Controllers of R & D be used as contractors who would compete for the right to supply the R & D or advisory service.

It can easily be imagined, against the background of the Public Service's burgeoning competence in technology assessment (roughly outlined in the first section of this chapter), what the general reaction to the CPRS's recommendations were. For illustration I append what were my own immediate thoughts.

> Large funding for relatively uninspired, improvized means of satisfying customer need with inferior technological solutions could increasingly occur if the Rothschild proposals are implemented.
> Despite Lord Rothschild's recommended 10% general research charge for the provision of marginal sustenance to "purer" research which might yield superior technological solutions to collective problems, fewer innovations could arise.
> Within an internal market for public service technology there could be a subconscious suppression of concern about damaging side effects stemming from proposed research if funding largely depends upon persuading relatively ascientific customers to foot the bill.
> Unavoidably, some moral responsibility is removed from the researcher

---

[1] My own investigations lead me to believe that the CPRS had not tried or bothered to make their study less insular by looking, even in a cursory fashion, at foreign practice in governmental control and surveillance of technology.

when a customer has given him an explicit commission to work in a specific field.

In order to maintain a necessary minimum of surveillance on methodology used in customer—contractor advocacy systems, the Chief Scientific Advisor's Office, within the Cabinet Office, would need to grow in size to such an extent that it might usurp departmental functions and cause an administrative spiralling growth at the centre of the administrative system.

No employee, within the constraints of the public service, can completely simulate a customer or a contractor when he is dealing with a fellow public servant. Effective role-playing is only truly possible when one of the customer—contractor pair is responsible to an organization outside the service. Lord Rothschild's suggestion that staff should, in some cases, work with competing loyalties is untenable. Careers are fashioned more by departments than central organizations, particularly at the working level where much detailed evaluation and assessment of technology occurs.

The major upshot of the Rothschild Report to date (June 1972) has been that the U.K. Parliament's Select Committee on Science and Technology took a very active and open hand in the resulting debate. The Committee, in democratic contrast to the way CPRS had compiled its flimsy evidence, took some pains over gathering, *in public*, opinion and evidence on "Government Involvement in and Policy Towards Research and Development". The "ogres in the scientific garden" (see Tucker, 1971) revealed by Lord Rothschild were examined by the Committee. Despite being handicapped by a lack of research backing that an Office of Technology Assessment, responsible only to Parliament, could have provided, the Select Committee were able to report conclusions which, according to Benn (1972), were "a most important political document which will have considerable repercussion in Whitehall, the scientific community, and the universities". In their report, the Committee criticized the Government[1] for prejudicing the Green Paper containing Rothschild's recommendations and intended only for discussion by an advance statement of policy accepting the CPRS proposals in principle. Furthermore, the Committee said the Green Paper contained

"no clear recommendations on the organizational means by which development should be carried out by the Government and applied to the national advantage...".

---

[1] The criticism of the Government in this matter by the Select Committee has enhanced significance because the Committee has pan-party membership.

And in one matter the reactionary criticism by the scientific establishment was said to be justified. The Committee's report suggested that Lord Rothschild had not allowed adequate consultation with the Research Councils and others (see Tucker, 1972). Of greatest significance to supporters of the concept of an Office of Technology assessment, however, was the Committee's suggestion that Lord Rothschild

> "gave such cursory attention to two of the most difficult problems for Government R & D, namely, determining which programmes are most worthwhile and what should be spent on them, and ensuring that the results are exploited to the greatest public benefit".

Because of departmental fragmentation and the lack of information reaching Parliament, adequate structure and machinery for the discussion and determination of scientific priorities and national policy were found by the Committee not to exist.

The Report also suggests that no action be taken on the Rothschild Report until a full investigation has been made on departmental research and development spending and priority determination of the current expenditure of $ 645.5 million per annum.

Other Select Committee recommendations of great relevance to the case in favour of a British Office of Technology Assessment are

> There should be a Minister of Research and Development with his own vote who should be a member of the Cabinet with statutory power to examine and approve all government R & D.
>
> A statutory Council for Science and Technology should be set up consisting of not less than 12 persons, with the Minister for R & D as Chairman. The council would advise on the formulation of priorities and policy for expenditure on civil and defence R & D, assess to likely or potential needs of the community in relation to developments in science and technology, anticipate and provide the necessary R & D to deal with any hazards which may threaten the community, and commission its own studies whereever necessary from departments, research councils, industry, or university groups.
>
> The Minister of Research and Development should make a full annual report to Parliament.
>
> All government departments with R & D activities should publish in a standard form, annual reports on these activities. The reports should include statistics explaining the size of their total R & D budget, progress reports on projects, assessment of the results of former R & D work, machinery for dialogue with potential users, customers, and contractors.

There is no need to compare point-by-point the similarities between the above findings and the advantages enumerated in Chapter 2 for the establishment of Offices of Technology Assessment (OTA's)[1]. The Rt. Hon. Wedgwood Benn has summed up the above debate in words which neatly summarize the feelings of most public and private assessors of technology (see Benn, 1972).

"When the dust has settled Lord Rothschild will be seen as having been the trigger mechanism for a process of democratization that goes far beyond the customer—contractor principle which caused such an uproar. Whatever the ultimate outcome of his report he will be remembered as the man who lifted the lid off the private world of decision making and did it in such vivid and abrasive language that a lot of other people will now be taking a great deal more interest in everything that happens there".

"Most significant of all is the greatly enhanced reputation of the Select Committee and the House of Commons itself which is bound to follow from this report. The Committee has only been in existence for a very few years. When it was first proposed, many voices were raised in Whitehall against its establishment. Some Ministers and senior Civil Servants were very hostile indeed to the idea of it probing into decisions that had traditionally been the preserve of experts. Though I strongly favoured the formation of the Committee, until I heard these arguments against it being deployed with real passion I never really understood the full importance of what was being done. Nor, until I appeared as its first Ministerial witness, did I appreciate the immense (and welcome) accretion of Parliament (as distinct from official) power that it represented. This was not to be the relatively easy contest in the ring at Question Time, with the Speaker stopping the fight after two supplementary punches, long before they began to hurt. This was the real thing, deep probing into policy and thinking with witnesses called to give in public evidence they might have previously submitted in private that would never have seen the light of day."

Conclusions

It could very well be that I have examined the situation and evidence produced by the recent catalytic episode (or debacle) in British politics and administration of technology with the eye-of-faith. Be that as it may, I cannot do other than report a stirring of parliamentary consciousness towards the concept of an Office of Technology Assessment. Perhaps after this has been written there

---

[1] My own pleasure in reading the Select Committee's suggestions for handling the problems, publicly aired by Lord Rothschild, can be imagined when it is realized that this tome was begun at least six months before the Committee's report was issued, and that the Clerks to the Select Committee had kindly accepted a personal memorandum which commented upon the Lord Rothschild Green Paper in the light of international development of the OTA concept.

will be a reassertion of the rights of the concealed decision maker and in-house assessors will be on the qui vive and better able to erect defences against outside critics who wish to make public gladiators out of gray eminences. Alternatively, we could be heading for an overswing in the opposite direction where we suffer a surfeit of public empiricism without finding a remedy to real and potential technological ills.

One way of ensuring a reasonable balance between the extremes of assessment by gray eminences and public quackery is for the public to take a keener interest and help nurture healthy change, before proponents of the status quo act to nullify the present constructive trend towards greater objectivity and openness in public decision making on matters technological. Some modification of the well-known tag might be in order — "countries get the assessors of technology they deserve".

References

Benn, A.W. (1972), "For Science, Open Government has Arrived", *New Scientist*, 54, No. 795, 314.

Galbraith, J.K. (1967). *The New Industrial State*, Boston, Houghton Mifflin, p. 12.

Tucker, A. (1971). "The Ogres in the Scientific Garden", *The Guardian*, 9th December, 1971.

Tucker, A. (1972). "Cabinet Minister Must Take Over Research....", *The Guardian*, 5th May, 1972.

CHAPTER 4

The British Programmes Analysis
Unit (PAU)

"Chaos umpire sits,
And by decision more embroils the fray
By which he reigns: next him higher arbiter
Chance governs all."

John Milton, *Paradise Lost*

Comments on the PAU

In the European context, a discussion of Britain's new Programmes Analysis Unit (PAU) is germane to any general discussion of technology assessment, particularly that part concerned with the evaluation of research and development. The writing of a chapter on the PAU may be a cathartic experience for this author (who served there), but it is hoped that even a subjective account of the Unit's past activities will also be of less personal and wider significance because of the catalytic effect the PAU has had on the practice and growth of technology assessment in the United Kingdom.

Many unsolicited testimonials have been made verifying the last assertion. However, for an ex-employee of the Unit to list approbations given to the PAU is unseemly and one commendation (made in the third year of the Unit's existence) must suffice to set

103

the scene. Hill Samuel and Co. Ltd., the merchant bankers, commissioned an independent study of "The role of Government in Research and Development" from Maxwell Stamp Associates Ltd. The outcome was Hill Samuel Occasional Paper No. 4 (1969)[1] which contained the following statement.

> "The newly created Programmes Analysis Unit is the only organization that so far exists for the intensive theoretical study of the problems attached to the economic implications of research and development activities .....
>
> We, therefore, recommend that the new central committee be equipped with a secretariat that is capable of such analysis, and specifically charged with attempting to evaluate the prospects offered by the programme of the research agencies and developing appropriate techniques......
>
> Such a secretariat could either be set up by the transfer of the current Programmes Analysis Unit to the central organization, or by the creation of a new unit."

This consultant's statement, from which the early efficacy of the PAU may be inferred, is undoubtedly from a sector of society legitimately concerned with special business interests. But if the words "and social" are added to "economic" the statement may be accepted, subject to personal reservations about the recommendations made.

The initiatives of the former Minister of Technology which led directly to the formation of the PAU have already been listed in Chapter 1. When the Unit was established in the Spring of 1967 it reported to Dr. J.B. Adams in his dual capacity as Special Adviser to the Ministry of Technology on Manpower and Resources and Member for Research on the Board of the Atomic Energy Authority. The Unit was appropriately sponsored, and staffed, jointly by the Ministry and the UKAEA A statement made in PAU (1971) lists the Official aims of the Unit, which have remained invariant since 1967, as

> "To explore the wider applicability of appraisal techniques to Civil research programmes, with the ultimate objective of improving the deployment of the U.K.'s R & D resources and maximizing the returns from Ministry and AEA investment in this area".

---

[1] Note the propensity of organizations with vested interests to commission studies on matters related to technology assessment. Note also that this study preceded the setting up of Lord Rothschild's think-tank in the Cabinet Office.

Itemized terms of reference given to the PAU were initially summarized as follows.

To examine and develop techniques for the evaluation of R & D programmes,
to apply these techniques to specific programmes as required by the Unit's sponsors, and
to pass the techniques on to the laboratories of the Ministry and the AEA when proven.

Research workers naturally greeted the possibility of usefully applying formal appraisal techniques with some reserve and a diminishing number still display scepticism. Nevertheless, despite initial philosophic doubts, the U.K. was the first country to enter this field at governmental level by the creation of a special "in-house" unit for the evaluation of possible benefits, disbenefits, and costs of civil research, though, of course, some major industrial enterprises had already made a practice of evaluating potential returns on R & D for some years[1].

From shortly after its inception to the present the Unit has existed with a staff of about twenty five graduates comprising chemists, physicists, chemical engineers, mathematicians, economists, mechanical and electrical engineers, generalists, and one barrister. Nearly all the staff come from a technical or scientific background (even the generalists have worked alongside scientists for many years) and consequently they have gained considerable professional experience of scientific organizations both at home and abroad. With such a composition, the PAU can rightly claim to be multi-disciplinary but individual members all showed a strong interest in the management of science before recruitment to the Unit. Which is not to say that they were all advocates of more science, some were already showing disillusionment with the received benefits and disbenefits of science long before the "ground-

---

[1] The rate of diffusion in industry of *quantitative* project selection procedures has been found to be relatively slow. See Mansfield *et al.* (1972) who showed that the proportion of a sample of industrial laboratories using a quantitative project selection method followed a slow logistic curve (see Chapter 7) given by

$$V(t) = [1 + 23 \exp(-0.22t)]^{-1}$$

where t is measured in years from 1949. Mansfield's equation suggests it takes about ten years for half the available companies to adopt a new quantitative project selection technique!

swell" of public opinion had moved in that direction. During the PAU's formative period the Unit's managers admit to deciding that

"a pragmatic approach should be adopted with the development of methodology being undertaken as a part of actual evaluation rather than in isolation" and "in this way it was hoped that any developments in methodology would be realistic and practical rather than idealized solutions to non-existent problems".

However, in making this decision the Unit was protected (in my opinion) from a descent into the pragmatism of the mediocre by curiosity-motivated and extra-curricula efforts of some of its more outre thinkers[1]. Acknowledgement of the earlier emphasis upon pragmatism and generous readjustment of the last assertion has recently been made in PAU (1971), where it is now conceded that

"sufficient experience has now been gained to permit some relaxation of this philosophy" (of pragmatism).

The lesson to be inferred is that no unit charged with making technology assessments should attempt to align its staff completely towards pragmatism. A sprinkling of "conceptual thinkers" seems to be necessary for the success of evaluation exercises. This is not surprising since a little thought will convince the reader that no less creativity is required for assessment of scientific and technological innovation and invention than it is for creation. Future impacts of technology are likely to be so unexpected by a wide consensus of pragmatists that useful insurance against failure to make accurate forecasts may be obtained by employing people of higher than average conceptual perception.

Some support can be given for the Unit's earlier decision to undertake construction of methodology only during actual evaluations because it is certainly true that each problem of technology assessment has its own idiosyncratic difficulties. Universal concepts of analysis are not common, thus analysts are often forced to custom-build a rationale for the solution for each problem. Notwithstanding this, it does need to be stressed again that these individually tailored solutions spring more from conceptual thought than pragmatism. We could say that pragmatism is only the "filter" through which the techniques of analysis are passed

---

[1] This sentence is not part of my personal catharsis. In penning it I have in mind colleagues who suffered the slings and arrows of their outrageously "practical" co-workers in order to make conceptual points which were later seen to be valid.

106

before their validity can be assessed. What begins as highly conceptual can, after it becomes familiar, become highly pragmatic. In numerical terms it is useful to note that the first Director of the Unit presided over staff who were not more than four-fifths in favour of absolute pragmatism.

Since its formation, the Unit has carried out a large number of technology assessments, mainly in the form of benefit—cost analyses containing "measured" economic and "inferred" social costs and benefits. On 20th November 1970, Mr. Laurance Reed asked the Secretary of State for Trade and Industry whether he would list the past and present studies undertaken by the PAU. The text of the Parliamentary Written Answer, House of Commons, is as follows.

Completed Studies

For Ministry of Technology
* Carbon fibres and other high-strength materials research.
  Marine technology, fishing.
  Marine technology, mining.
† Marine technology, dredging.
  Cryogenics.
† Fire Research Station.
* Superconductivity.
  Aspects of the U.K. Space Programme.
  Drawing measuring machine.
  Cranes.
† Educational technology, computers.
* Hydrostatic extrusion.
  Industrial aerodynamics.
  Warren Spring Mineral Processing Group.
  Computer demand survey.
† Air pollution costs and research.
  In-process gauging.
  Mintech computer installation.
*† Nuclear ships.
  TV market forecast.
  Hovercraft and hydrofoils.
  Preforming of metals (interim report).
  Survey of computer services.
  Computer demand survey, Europe.

Cosmetic dentistry.
Garment making machinery.

For the UKAEA
  *Desalination.
  High-temperature corrosion.
  Nuclear research.
  Metal working fluids for power.
  Magnetic materials (ferrites).
  Ceramic heat exchanger material.
  Graphite electrodes.
  Siltation studies.
  Industrial application of radiation chemistry.
    (a) Paint curing.
    (b) Food sterilization.
    (c) Radiation-induced chemical reaction.
  Ground and river waters.
  Ceramic-tipped tools.
  Electrohydraulic crushing.
  High-temperature fuel cells.
  Industrial physics (isotopes).
  Ion implantation.
  Internal combustion steam turbine.
  Non-Destructive Testing Centre.
  Glow discharge heating.
  Heat pumps.
  Condensers/non-condensibles.
  Low-power energy sources.
  Welding machine.
  Neutron source reactor.
 †Advanced transportation systems.

* Study commissioned jointly by Ministry of Technology and the UKAEA
and/or by agreement reported to UKAEA in addition to the Ministry.
† Many studies involve collaboration with various Government Departments.
Items marked † were specifically reported additionally (through Ministry of
Technology) to other Government Departments/Agencies.

Current Studies (1970)

For the Department of Trade and Industry
  Electro-heat processes.

Pattern cognition and recognition.
Costs of defective materials.
Benefit of computers in R & D.
*Fuel cells and secondary batteries.
Diffraction gratings.
Surface finishing.
*Carbon fibre research, review.
Future demand for machine tools.
Joint fire research.
Organisation, capital facility.
Basic research, X-rays.
Torry research ship (fish processing).
Industrial aerodynamics, review of needs.
*Hydrostatic extrusion, reappraisal.

For the UKAEA
Applied nuclear research.
Analytical R & D unit.
Basic nuclear research review.
Desalination review.
Tribology.
High-temperature chemical review.
Ion implantation review.
Power fluidics.
Water renovation.
Ceramics, technological forecast.

* Study commissioned jointly by the Department of Trade and Industry and the UKAEA in addition to the Department.

A perusal of this list, even for those who understandably find it difficult to comprehend completely what lies behind the brief titles of the different evaluations, will go some way towards convincing the reader that the PAU's activities were, and are, energetically pursued and dispersed over a wide field of interest. Naturally, the spectrum of R & D proposals evaluated by the PAU has widened since the above Parliamentary answer was made and not all current evaluations can be listed for sensible reasons of commercial security. There is, however, a general categorization which can be made about two main types of evaluation involved.

Singular projects concerned with a relatively narrow range of scientific or technological activity directed towards an end-product (or result) which should yield an economic or social benefit, and *sectorial projects* concerned with broadly important areas of science or technology which could have strategic or tactical importance to the achievement of economic or social benefit (or, equally important, the avoidance of disbenefits).

Other categories which arise, though only constituting a component part of technology assessment, are market surveys to ascertain the elasticity of demand for a specific technological good or service, and review of resource allocations made within laboratories.

Examples of "singular" evaluation projects from the above list are cranes, drawing measuring machine, ceramic-tipped tools, welding machine, etc.

Examples of "sectorial" evaluation projects are educational technology, e.g. computer-aided instruction, air pollution, marine technology, etc. Of greatest general concern to society are the sectorial evaluations which, ideally, should be the logical first stage assessment made *before* resources are channelled into new all-embracing fields of technological activity. Historically "sectorial" evaluations have not predominated over "singular" evaluations and the latter have tended to precede decisions about investment of resources. Only in rare cases, like the NASA Space Programme and National Atomic Energy Programmes, have resources been committed in the belief that an optimal selection of worthwhile sectorial activities has been made.

All decisions based upon singular evaluations are necessarily sub-optimal and there is no known law-of-nature which suggests that a mix of sub-optimal decisions will lead to an overall benefit (or not lead to an overall disbenefit). There are, of course, compatible sub-optimal decisions like those of two neighbouring communities one of which decides to fish in crocodile-infested waters with waders because they have no better technology than an efficient crocodile-jaw clamp (which must be hand placed) and the other makes wooden legs because it has carried out a market survey and sees an outlet for its wood-working technology[1]. The two operations are indeed compatible but not optimal in the total sense. A better solution for the collective benefit of these isolated communities would be that the wood workers should adapt their technology to telechiric construction of "pantographic" means of placing the crocodile-jaw clamps by remote control and that the

---

[1] I am grateful to A. Jones for prompting me to think about the dilemma of the two communities, see Jones (1972).

110

fishing community should work out some use for crocodile skins and carcasses.

In the absence of a truly cohesive framework, sub-optimal decisions are arguably better than nothing and, within the present constraints on central planning, "singular" evaluations proceed in PAU somewhat as follows.

Staff, given a remit to evaluate a specific project, embark upon an intensive programme to familiarize themselves with the range of scientific/technological background involved, often using the proposers of the project as tutors; they would then try to indicate (see PAU, 1971):
the worthwhileness of the project as a whole and its ability to recoup a reasonable return on investment (not necessarily to the sponsors):
the technical and commercial points of particular importance to benefit realization so that appropriate stress can be given to these in order to maximize the chances of full exploitation,
the significance of timing to the commercial success or failure of the project in order to avoid investing "too little too late" or "too much too soon", and
the economic and technical external factors which could materially alter the justification of the work.

In general, single-project evaluations result in recommendations concerning objectives, investment levels, time scale, technical content, and review criteria which are usually based on technical progress or identifiable externalities rather than time. In practice, PAU (1971) claims

"it has been found that the expected net returns (i.e. net of a further investment needed to exploit the research) on R & D investment are usually in excess of three times and frequently over five times the nominal R & D resource cost, including overheads, a finding which conforms with the estimates of several major industrial companies".

This last statement may be taken at its face value. However, it should be borne in mind that R & D investment comprises intellectual as well as financial investment and the "rules" for apportionment of overall benefit to the different parts of the chain linking laboratory-created innovation to the commercial market (or some other mechanism for the distribution of collective benefit) are only vague (see Chapter 8 where the notional apportionment of benefit is discussed).

When ascertaining the technical and commercial points of particular importance to benefit, the analysts must undergo a wide interaction with all interested parties (researchers, industrialists, academics, and potential customers). This process of interaction and information seeking can be fatiguing when coupled with the

111

short term nature of "singular" evaluations (one analyst could be asked to carry out many evaluations of very disparate projects in a year). The analyst suffers the usual exigencies of the "travelling man" and the continual need of becoming an "overnight" expert in new technologies. Signs of fatigue can be detected when the analyst puts more emphasis on visiting contacts than he does on acquiring a crucial understanding of the technology under review, or, conversely, puts more emphasis on desk-work and neglects useful contacts. Between these two extremes, known idiomatically by the deprecations of "commercial traveller approach" and "theoretical hand-waving approach", there is obviously a best approach which varies according to the ability of the analyst and the nature of the work to be accomplished. The judgement of what the optimal approach is can be a very fine art indeed and is only acquired after much experience of technology assessment. Throughput of PAU staff has not been as great as one would have expected when allowing for the exigencies of the work, and a possible reason might be that a "tired", but often very experienced analyst is preferable to a newly recruited energetic analyst who has only a small accumulation of experience in assessment.

Sectorial evaluation projects have taught the PAU some invaluable lessons which are of more relevance to the topic of technology assessment, when it deals with the impact on society of new technology, than the experience acquired in "singular" evaluation projects. In many ways "sectorial" work is less fatiguing than "singular" evaluation procedure, for example

> the analyst, having climbed up a "learning curve" peculiar to the particular sector, has a longer time to utilise this hard-won knowledge, and sectorial problems demand a greater mix of multi-disciplinary effort, so that the probability of generating a useful attack on the problem is increased [1].

Against these advantages should be set the generality of the amorphous remit that triggers the sectorial evaluation. Sometimes the remit is so bereft of guidance as to simply state: "What should be done in this sector?". Obviously "sectorial" analysis is more demanding because it requires greater creativity from analysts. The general PAU approach to sectorial problems has been one of struc-

---

[1] One would expect the number of useful interactions between n analysts to be not less than about $\frac{1}{2}n(n-1)$ and although no threshold of usefulness in larger numbers has been quantified (or lower-bound of confusion defined when the numbers become too large) the effect has certainly been qualitatively experienced.

turing the problems using mathematical modelling techniques and exploring the implications of different courses of government action with the aid of these models (see, for example, Hill, 1969, Medford, 1969a and b). The output of the PAU's sectorial evaluations has reached a wider audience than that comprised of the Unit's direct sponsors because other Departments with relevant statutory responsibilities must be kept informed, as well as the DTI and UKAEA who may only carry responsibility for the R & D and the provision of the means by which other Government departments implement their policies. Furthermore, sectorial studies have so wide a potential interest that it has always been official policy to let bodies outside the public service see them, when there are no restrictions of commercial or other security [1]. This last democratic procedure has not yet been formalized and there is no method of making it illegal *not* to give a sectorial analysis *wide* publicity. Before the emergence of the American concept of an Office of Technology Assessment the U.K. Government was second-to-none in democratic procedure for granting permission for external publication of in-house studies. When the American Office of Technology Assessment becomes operational (see Chapter 2) the U.K. will have fallen far behind since the original Bill H.R. 10243 contained the following clauses, aimed at providing more public information.

"For this purpose the Office is authorized to require the attendance of such persons and the production of such books, records, documents, or data, by subpena or otherwise, and to take such testimony and records, as it deems necessary. Subpenas may be issued by the Director or by any other person designated by him" [2] and

"The Office shall submit to the Congress an annual report which shall, among other things, evaluate the existing state of the art with regard to technology assessment techniques and forecast, insofar as may be feasible, technological areas requiring future attention".

Demand forecasts, or market surveys which fall into neither of the last two categories discussed, have increasingly become part of PAU's work. Apropos of these it may be salutary to say something

---

[1] This is well illustrated by the list of openly published reports and papers by PAU authors on sectorial and singular evaluation procedures appended to this chapter.

[2] This clause from the first version of HR 10243 is no longer explicit in the version of 9th February 1972, because of the Brookes' Amendment (see Chapter 2) but the same subpena powers may still be inferred from the Congressional membership of the OTA Board.

of the dangers of unbalanced evaluation procedures although it should not be understood that the following remarks are directly aimed at the system within which the Unit presently operates.

Reaction of sponsoring bodies or some section of a sponsoring organization to an evaluation unit may be predicted from the Principle of le Chatelier, which states that

> "If to a system in equilibrium a constraint is applied, the system tends to adjust itself so as to oppose the constraint and restore equilibrium."

One likely gambit through which an apprehensive sponsor could attempt to restore any desired equilibrium that an independent analysis might threaten would be to use the evaluation unit for the provision of only a partial service. For example, the sponsor could try to utilize the unit's expertise in projecting market demands: thus the sponsor is *seen* to be utilizing the service provided *but* only for a market research survey.

Such a utilization is, of course, unsatisfactory because all market surveys should be accompanied by a realistic assessment of whether the product can be produced in the form assumed during market analysis and, if it can, whether (regardless of any possible lack of discrimination by the users) the use is a good application of the product.

Naturally, one assumes that any such requests for the limited use of evaluation procedures are to be supported by complimentary work of other analysts which will make good these deficiencies.

Any noticeable imbalance in overall evaluation procedures which puts too much evidence upon market surveys, without there being tangible evidence that the product is manufacturable or is the optimal way of currently satisfying customer need, should give cause for concern about the possible use of analysis for the advocacy of a second-rate programme[1].

The External Contacts Made by the Unit

An organization acting within the Government service has natural advantages when carrying out technology assessment. It can

---

[1] This statement is made despite the author's appreciation of the professional efforts of many market researchers and his willing involvement in the professional activity of industrial marketing research. Market research as a functional aid is not under attack in the above sentence.

receive invaluable and willing assistance from industrial enterprises, universities, and government laboratories because it has not been categorized as a competitor. Even other governments frequently lend assistance on large sectorial problems of common interest when there could, conceivably, be some excuse for regarding another government's assessors as precurrers [1] of competitive threat. The government analyst is, therefore, granted advantages that his industrial colleague cannot lay claim to and is forced to seek through less direct approaches like reiterative questionnaires (see, for example, the Delphi technique [2] discussed in Chapter 7).

Naturally the government analyst is given much information in strict confidence and it is not usually feasible to reproduce overt versions of this information. Nevertheless, by carrying out a series of recurrent interviews it is possible circumspectly to cross-fertilize and check opinion given in confidence. Such a procedure is often referred to as a "two-legged Delphi".

Freedom of access to other analysts and organizations interested in the application of technology assessment generates return requests for specific exchanges on evaluation and assessment technique. Such requests have been acceded to and the Unit, in fulfilling its remit to pass on techniques, has given invited contributions to conferences or meetings sponsored by the British Institute of Management, The Institute of Physics and the Physical Society, the Institute of Electrical Engineers, the Society for Chemistry and Industry, The Institute of Chemical Engineers, The Operational Research Society, The Institute of Patentees and Inventors, The Industrial Marketing Research Association, The International Institute of Public Finance, The European Industrial Marketing Research Association, The European Technological Forecasting Association, the Universities of Aston, Bradford, Edinburgh, and Oxford, the Manchester and London Business Schools, NATO, OECD, The Committee of Directors of Research Associations, The Civil Service College, and the U.K. Treasury Management Accounting Unit amongst others.

The PAU's liaison activities have not been restricted to the United Kingdom. Close contact on methodology and philosophy

---

[1] Shakespeare Ph. and Tur 6, forerunner, precursor.

[2] The Delphi questionnaire technique has natural advantages when information and opinion is not freely given in (say), committees but it is very noticeable that the technique is favoured more outside, than within, Government service.

has been maintained with others active in the evaluation and fore-casting field particularly in the U.S.A. and France. Some measure of international interest may be gauged from the fact that the PAU has been officially visited by representatives from no less than sixteen nations.

The Unit's role in promulgating knowledge of technology assessment has not been acquired because it is the leading exponent of the art in the U.K. (although the author would maintain that the Unit ranks equal first, in this respect, with a few other organizations). It is because the Unit occupies an informal, central, coordinating role, without executive power to do other than recommend technique and does not need to regard expertise in technology assessment as a commercial asset.

An Office of Technology Assessment responsible only to Parliament would, of course, exercise far greater impact upon the use and generation of assessment techniques because given legislation like that in HR 10243 it would in "present(ing) findings of completed analyses to the appropriate legislative authorities", and "making freely available to the public" information, surveys, studies, reports, and findings, be able to act even more cohesively in this respect.

The major outcome of the Unit's pioneering existence could, given a willingness to legislate for a European or British OTA, be that it has helped create a favourable climate for the acceptance of such legislation and that it has also provided a number of experienced analysts who would almost certainly be willing to support such an Office.

Attitudes of the Unit's Staff to Their Past Experience

It is not for this author to claim too accurate a discernment of the reaction of his past PAU colleagues to their collective 150 man-years (or so) of technology assessment experience within the Unit. Too close an involvement in the experience naturally precludes an objective assessment of other peoples' reaction to the stimulus of living check-by-jowl with the need to make assessments over a period of some five or six years. However, some highly subjective interpretations of colleague's reactions will be given in the hope that they will be useful to a general understanding of the human factors involved.

There is a discernable feeling that a large number of evaluations have

116

been too hurried and made in the absence of sufficient data (this needs to be balanced with the real exigencies of decision making where the need to make a decision can be paramount).

There is a decided view that many dubious techniques of analysis have been foisted upon the sponsors of analysis.

There is a reluctance amongst many members to admit that there has been any great reliance on technological forecasting[1].

Natural rebuttal of the first reaction has been made. Rebuttal of the second must rest on the admission that, during a learning period, mistakes in methodology will be made and most of what is not directly familiar must initially be suspect. The third reaction does not match the facts because nearly all members of the Unit have carried out technological forecasts falling within Jantsch's definition of "probabilistic assessment, on a relatively high confidence level, of future technological transfer". The denial by some individuals[2] of participation in forecasting and reluctance to publicize examples stemmed, in the author's opinion, from a healthy scepticism about the product and the belief that technological forecasting was not worthy of the status of a discipline, despite the fact that forecasts had been produced by a ménage à trois (between science, art, and experience) in the face of inadequate data and the need for a prognostication on which to base an imminent decision. Most of the Unit feel that there is nearly always inadequate data at the time a forecast is needed regardless of their industrious professional efforts to garner and collate the relevant facts. It was also possible to detect a residual feeling that the logical insecurities, imagined to be in the forecast, were a consequence of personal inadequacy and not the inherent difficulty of all prognosis.

Another major bone of contention is that the analysts do not always know how their analyses have been used. Sometimes the assessment is submitted to a "steering committee" who may or may not reveal to the analysts what their reaction is. On other occasions the process is not so clear. Sometimes there is useful individual communication from arbiters-of-decision who have used the assessment. Occasionally the analysis disappears into an administrative maw and a decision can emerge which is apparently incompatible with the tendered analysis. Perhaps this is as it

---

[1] See Chapter 7 for a description of technological forecasting.

[2] As a collective body the Unit emphasises its participation in technological forecasting and certain individuals (the author included) are decidedly not reticent about their activities in this field.

should be because arbiters are not bound to keep their analysts informed of the details of the decision making. But frequently the analyst becomes sensitive to the way his work *might* have been misused and there is a good case for keeping analysts more closely informed of the mechanism of the decision process. The most comforting situation which does occasionally occur in the Unit is that in which the analyst clearly sees what decisions have been taken following an earlier assessment and he is asked to re-assess the situation in the light of the consequences that have followed the initial decision. Such a continuing tie between the arbiter and the analyst replaces some of the lack of managerial responsibility felt by the analyst (when he has been recruited from a relatively senior position that entailed executive responsibility).

All-in-all analysts tend, because of the natural exigencies of their jobs, to be excessively critical of the work done by themselves and colleagues. Strangely, the rationalization that assessment, carried out in the presence of so much uncertainty, must of necessity lack much in rigour does not carry much solace. Conversely, it tends, in reality, to exacerbate the dissatisfaction. The only cure for too much pessimism is for the analyst to be brought face-to-face with the consequences of decision making which is made without the support of technology assessment.

A Distillation of Lessons Learnt by the Unit

The lessons of the first four years of the Unit's existence have been officially given in PAU (1971). This section deals with a paraphrase of these interwoven with the author's own interpretation of the lessons learnt.

Programme evaluation and assessment has, to a large degree, now been accepted as an integral, though by no means all-embracing part, of the decision-making process. This acceptance has occurred during a period in which a noticeable change has taken place in the attitudes of applied scientists in the U.K. and in which public opinion, as manifest by the transfer of university enrolments from science to arts-based courses[1], has exerted some pres-

---

[1] School Leavers, CSE and GCE, HMSO (1970) and (1972), suggests that swing away from science subjects in school leaving examinations levelled out in 1967 and 1968 and the movement back accelerated in 1970. See also *The Guardian* 20 Jan. 1972.

sure on the scientific "establishment". There is now a far greater awareness on the part of scientists who administer public investment in R & D of the need to relate programmes of work to some realistic benefit to Society. Concurrent with this there is also an abandonment of an old attitude "that the exploitation of R & D results is a task better left to the lower echelons of the intellectual elite"[1]. At a time when science is under attack, albeit sometimes an uninformed one, the scientific community has become aware that it must be seen to be doing good and not mischief. It is thought that the activities of PAU have, within Government service and possibly wider spheres, acted as a catalyst in this process of change.

On the other hand, the marginal activities of the Unit in areas related to pure, or curiosity-motivated, research (see Chapter 8) have acted as a restraint on the application of inappropriate evaluation techniques to areas where the possible outcome of acquiring new knowledge can barely be perceived.

The paramount importance of personal contact with all parties interested in the outcome of applied research programmes, i.e. enquiry (or intelligence) into the social and industrial aspects of new innovation, has been clearly established. Because "intelligence" has emotive overtones it should be regarded in this context as the intelligent collection and collation of pertinent information for civil purposes only.

The direct value of analysis is prima facie obvious but additional benefits have accrued which are not, at first sight, apparent. The interaction between the PAU and applied scientists has led to a more precise definition of objectives, sometimes even when programmes are already underway with goals that appear to be fixed and inflexible. It has been argued, see PAU (1971), that "the analyst can devote more time than laboratory scientists to exchange with industry and this in many instances has materially increased the chances of successful exploitation and acted as a general stimulant to innovation by helping to bridge the gap between the scientist and producer or user". In relation to this last remark, I would like to remind the reader of the possibility that *partial* analysis in focussing attention onto *some* of the essentials, can lead to the maintenance of funding for unfruitful projects.

The problems which occur in the management of R & D and

---

[1] This is *not* a quotation from the PAU. To avoid justifiable, but wasteful, verbal reprisals the source is left anonymous.

technological innovation are myriad and it could be that the Unit's painstaking critical (and creative) efforts in the methodological field of analysis will ultimately give rise to the greatest benefit obtained from the first half-decade of its existence. Despite detection of much that is dross in the new methodology thrown up by management "science" enough evidence has been accumulated to establish the need for such techniques. The efforts of past and present members of the Unit ensures continuance of an operable "critical facility" even in these days of the "hard sell" by vendors of new techniques. The different stances of politicians (and administrators) towards techniques which promise a rationalist approach to government policy making need to be buttressed by a clearer understanding of the "nuts and bolts" which underpin some of the imported (and indigenous) apparently rational, scientific, methods of management currently receiving belated attention. Nowhere near enough appreciation exists of the effort and time required to detect the hidden flaws, some of which can be serious, in some "rational" methods (like relevance techniques, planning programming and budgeting, etc.). Without the aid to critical discernment, provided by an experienced organization, there is a great danger that useless sophistry will be unknowingly applied with the very best of intentions.

PAU very often appears to be "hoist by its own petard" when it is asked the ubiquitous question: "What benefit, or benefit—cost, has the Unit achieved from its own activities?". Such a question is not definitively answered because the outcome of PAU evaluations, even when proffered advice has been followed, is not easily compared with what would have happened if no technology assessment had been made. It is, of course, known within the Unit what directly quantifiable benefits have been achieved when programmes of R & D have been rationalized (diverse efforts coordinated, unnecesarily competitive, parallel, efforts coalesced, etc.). These alone, in my opinion, give acceptable lower bounds of benefit and benefit—cost ratio. The Unit's continuing existence since 1967 and the official attempts to propagate its functions gives some qualitative feel for the value that governmental arbiters-of-decision place upon its benefit to them. If these arbiters take advantage of the "firm foundation" which "has been constructed on which to base future studies of a more complex and wide-ranging kind, e.g. inter-sector priorities" (see PAU, 1971) it will be clear that they attribute benefits not only to the Unit as a working service group but also as a group capable of methodological innovation in administration.

120

Conversely, if no attempts is made to "stretch" the PAU and the unit is encouraged to ossify into a "service group" which provides partial assessments, we shall know that the departmental system has reestablished equilibrium after a permanent, incremental, change for the better.

References

Hill, K.M. (1969), "Technological Forecasting in some Basic Industries", in *Proc. 9th Commonwealth Mining and Metallurgical Congress*, Vol. 1, The Institute of Mining and Metallurgy, London.

Hill Samuel Occasional Paper No. 4 (1969). "The Role of Government in Research and Development", Hill Samuel and Co. Ltd., London.

Jones, A. (1972). "Industrial Marketing Research in Developing Countries", *Industrial Marketing Management*, 1 No. 2.

Mansfield, E. *et al.* (1972). *Research and Innovation in the Modern Corporation*, London: The MacMillan Press Ltd.,

Medford, R.D. (1969a). "The New Thaumaturgy of Governmental Research and Development", *Futures*, 1, No. 6.

Medford, R.D. (1969b). "Marine Mining in Britain", *Mining Magazine*, 121, Nos. 5 and 6.

PAU (1971). *The Programmes Analysis Unit: 1967—1971*, PAU, Chilton, Didcot, England.

Appendix. Published Reports and Papers by PAU Authors

PAU Memoranda

1 PAU.M.3/67	The analysis of research and development in the Ministry of Technology, by K.G.H. Binning.
2 PAU.M.4/67	Technological forecasting and the planning of research and development, by D.R. Coates.
3 PAU.M.5/67	Programme evaluation in the Programmes Analysis Unit, by R.L.R. Nicholson.
4 PAU.M.6	Some remarks on the application of technological forecasting, by R.D. Medford. (See next section for further details).
5 PAU.M.7	Technological forecasting as a management technique, by R.L.R. Nicholson. (Available from H.M.S.O., 20p., SBN 11—980420—4).
6 PAU.M.8	Project evaluation — Decisions on levels of investment in research, feasibility studies and market surveys, by P.M.S. Jones. (Available from H.M.S.O., 12½p., SBN 11—980529—4).
7 PAU.M.9	How much basic research is enough? (See next section for further details).

8 PAU.M.10	Technological forecasting as a management tool, by P.M.S. Jones. (H.M.S.O., 25p., SBN 11—980822—6).
9 PAU.M.11	Evaluating civil research and development projects in the Government sector, by R.L.R. Nicholson. (H.M.S.O., 9p., SBN 11—980830—7).
10 PAU.M.12	An outline of evaluation as practised by the Programmes Analysis Unit with three case studies, by P.M.S. Jones and H. Hunt. (H.M.S.O., 60p., SBN 11—980831—5).
11 PAU.M.13	Operational research in research and development, by P.S.S.F. Marsden. (H.M.S.O., 15p., SBN 11—980883—8).
12 PAU.M.14	Ranking development options, by P.M.S. Jones. (H.M.S.O., 12½p., SBN 11—981153—7).
13 PAU.M.15	Further remarks on forecasting techniques used in PAU by R.D. Medford.
14 PAU.M.16	Preliminary thoughts on social forecasting, by P.M.S. Jones. (H.M.S.O., 15p., SBN 11—981154—5).
15 PAU.M.17	The use of cost—benefit analysis as an aid to allocating Government resources for research and development, by P.M.S. Jones.

Other Publications

16 The Programmes Analysis Unit, by K.G.H. Binning, *New Technology*, No. 16, April 1968.

17 Technological forecasting in some basic industries, by K.M. Hill, in *Proc. 9th Commonwealth Mining and Metallurgical Congress*, Vol. 1. The Institution of Mining and Metallurgy, London, 1970.

18 How much basic research is enough? A problem of resource allocation, by K.M. Hill, L.G. Brookes and H. Hunt. *J. Long Range Planning*, Vol. 1, No. 3, March, 1969.

19 The uncertainties of planning major research and development, by K.G.H. Binning. *J. Long Range Planning*, Vol. 1, No. 4, June 1969.

20 Illustration of technological forecasting for R & D evaluations, by T. Garrett, in *Technological Forecasting and Corporate Strategy*, G. Wills, D. Ashton, B. Taylor, eds. Bradford University Press and Crosby Lockwood and Son, London, 1969.

21 Some remarks on the application of technological forecasting, by R.D. Medford, in *Technological Forecasting*. R.V. Arnfield, ed. Edinburgh University Press, Edinburgh, 1969.

22 Forecasting the need for research and development, by H. Hunt. *Futures*, Vol. 1, No. 5, September, 1969.

23 Marine mining in Britain, by R.D. Medford. *Mining Magazine*, Vol. 121, No. 5, November, 1969, No. 6, December, 1969.

24 The application of the credibility concept to research projects, by P.M.S. Jones. *Operational Research Quarterly* (letter), Vol. 20, No. 4, December, 1969.

25 The new thaumaturgy of governmental research and development?, by R.D. Medford. *Futures*, Vol. 1, No. 6, December, 1969.

26 *Computer Based Learning Systems: A Programme for Research and De-*

velopment, by K.M. Hill, R.D. Medford, P. McLaren and others. Available from National Council for Educational Technology, 160 Great Portland Street, London, W.1.

27 Investment in innovation, by P.M.S. Jones, *I.E.E. Conference publication No. 61*, pp. 181—195. Institution of Electrical Engineers, London, March, 1970.

28 Planning and managing the R & D programme, by P.McLaren, in *Proceedings of a Seminar on Computer Based Learning Systems*, Leeds University, September, 1969. J. Arnett, J. Duke, eds. National Council for Educational Technology, London, 1970.

29 The selection of R & D programmes with particular reference to new materials, by P.M.S. Jones, in *New Horizons for Chemistry and Industry in the 1990's. Proceedings of a Symposium held at Lancaster, July, 1969*. Society of Chemical Industry, London, 1970.

30 An example of R & D programme evaluation, by T. Garrett, *Fact*, No. 78, p. 13, June, 1970.

31 Technological forecasting as a stimulus to innovation, by R.L.R. Nicholson. *The Inventor, J. Inst. of Patentees & Inventors*, Vol. 10, No. 3, p. 22, September, 1970.

32 Economic considerations relevant to the development of new materials by P.M.S. Jones. *J. Materials Science*, Vol. 5, No. 9, p. 796, September, 1970.

33 Determining priorities and investment levels in scientific research and development, by P.M.S. Jones. *Policy Sciences*, Vol. 1, p. 299, December, 1970.

34 A note on some economic considerations for Government policy towards Hovercraft and Hydrofoils, by K.M. Hill. *Hovering Craft and Hydrofoil*, Vol. 10, No. 3, p. 14—15, December, 1970.

35 The practical application of cost—benefit analysis to R & D investment decision, by R.L.R. Nicholson. To be published in the *Proceedings of N.A.T.O. Symposium on Cost—Benefit Analysis*, The Hague, July, 1969, by English Universities Press, March, 1971.

36 Evaluation of Government research and development programmes, by P.M.S. Jones. *R & D Management*, Vol. 1, No. 2, February, 1971.

37 Lessons from the objective appraisal of programmes at the national level — implications of criteria and policy, by P.M.S. Jones. To be published in *Research Policy*, Vol. 1, No. 1, 1971.

38 Market Research in the novel product field, by P.M.S. Jones. *IMRA Journal*, Vol. 7, No. 1, February, 1971.

39 Market Research and R & D priorities, by R.L.R. Nicholson. To be published in *Proceedings of the Bradford Conference on Marketing Advanced Technology*, May, 1971.

CHAPTER 5

Benefit—Cost Analysis in Technology Assessment

"Contrariwise, continued Tweedledee,
if it was so, it might be; and
if it were so, it would be; but
as it isn't, it ain't. That's logic."
Lewis Carroll

Introduction

Benefit—cost analysis, as used in technology assessment, is basically an importation from the U.S.A. because the method owes much to the pioneering efforts made in the analysis of large-scale civil engineering projects since 1902 by the U.S. Army Corps of Engineers[1]. Prest and Turvey (1965) give a comprehensive and comprehensible survey of development of the subject prior to the European boom in technology assessment, but European readers will be pleased to see that they acknowledge the early pioneering work of the French economist Dupuit (1844).

---

[1] Since this book contains a plea for European and international legislation on Technology Assessment it is apposite to point out that the Corps' efforts on benefit—cost date to the Rivers and Harbours Act of 1902 which required the submission of reports on commercial benefits and costs of the Corps' projects.

It is not presumed that this chapter shall deal exhaustively with the known theory of benefit—cost analysis which is a very wide subject indeed. Only facets of benefit—cost analysis relevant to the assessment of social or market impact of technology will be dealt with here.

The purpose of benefit—cost analysis may be narrowly taken to be the quantification and sometimes the optimization of an arithmetic combination of expected-total-benefit and expected-total-cost of a project, where the "totals" are summations over *all* present values of costs and benefits which the analyst believes to be pertinent to the defined project and which lie within the scope of the analysis defined by his masters (usually the ultimate arbiters-of-decision, or their nominated intermediaries). Choice of the arithmetic combination to be used by the analyst is usually idiosyncratic to the organization involved or the nature of the programme under analysis. But the two most favoured arithmetic combinations are

(*1*) the quotient: expected-total-benefit divided by expected-total-cost, and

(*2*) the expected-total-benefit net of expected-total-cost.

In practise, this last narrow (arithmetically based) viewpoint is not adhered to by the analyst who often takes benefit—cost analysis to be a means of carrying out sensitivity analysis, i.e. the estimation of changes in the combination or any of its constituent component parts consequent upon changes of the input variables to the analysis, like time, probability of success, discount rate, etc.

In fact, benefit—cost analysis has really become a new media for communication between the arbiters and the assessors rather than a purely quantitative tool of assessment. Under the strict scrutiny of benefit—cost analysis, projects are very often revealed, as initially defined, to be too amorphous. The analyst, before beginning effective quantification, must then seek repeated redefinition of the project, and the background against which it is to be analysed, from the ultimate arbiters-of-decision. Given tolerance on both sides and an appreciation of the psychology of analyst—arbiter relationships, this iterative prelude to analysis then serves as an essential channel for necessary communications.

Analyst—Arbiter Relationships in Benefit—Cost Analysis

The search for adequate project definition can be a punitive exercise for analysts who are faced with masters who are unreasonably assertive and see any recurrent request for clarification as an attempt to usurp the higher functions of decision. Even tolerant masters become frustrated when, despite fairly tactful behaviour on both sides, it is made manifest that the delegation of instruction is, within the new analytical context, loose and unstructured. Sad to relate, under these circumstances some analysts, rather than risk a pyrrhic victory for objectivity, do covertly usurp the ultimate decision makers perogative or, even worse, stoop to obscurantism!

Arbiters-of-decision naturally face equally distressing experiences when participating in this demanding field of analysis. They can be beset with unimaginative analysts who regard themselves as glorified accountants and demand excessive definition of projects without being willing, or able, to make their own creative contribution to the problems of analysis. Another frustration for the arbiter is to be assisted by analysts who are only interested in obtaining intellectual stimulus from conceptual problems and have no patience with the less esoteric, but necessary, parts of the analysis.

Enough has been said to indicate that communication between arbiter and analyst is not easy and if it breaks down, with conflict pre-empting communication, changes in relationships are advisable. The arbiter can normally depose his analysts if he perceives irremedial conflict. But the remedy for the analysts is not so simple. Analysts still in conflict with arbiters, after a reasonable dialogue has taken place, may face problems of integrity, even after being relieved of direct responsibility for part of the project analysis. This is particularly true of benefit—cost analysis involving political value judgements, particularly those analyses carried out within the public service. The alleged recent behaviour of the ex-RAND employee, Daniel Ellsberg, who is said to have made the "Pentagon Papers" available to the New York Times, is a case in point and could very well become a classic case of the behaviour of analysts under stress.

There is, in my opinion, a whole field of study of the psychology of arbiter—analyst relationship which could well repay investigation. No worthy arbiter wants a "tame" analyst because this would suppress constructive conflict, but at what level of conflict should an analyst be put to other work? And has the analyst any right to publicize his disagreement with his masters after being relieved of direct responsibility?

These asides to the main business of describing cogent parts of benefit—cost analysis technique have been made at the onset of this chapter *not* to exacerbate unreasonable ambition on the part of analysts. They have been made simply to stress the need for good relationships between analysts and arbiters in the exercise of analysis. No matter how versatile an analyst may be, he will not be able to operate effectively unless he can acquire some skill in working honestly, but not disruptively, with decision makers. Similarly, decision makers should realize that in employing benefit—cost analysts they are committing themselves to the need for more detailed explanation of their thoughts and plans, than is customary in normal decision maker—staff relationships.

Present Value

Since the concept of present value is deeply embedded in the methodology of benefit—cost analysis, it is appropriate to consider the subject in some detail.

The use of present value technique has significance because it transforms components in a benefit—cost relationship from functions of *time*, money, probability, and other quantities into functions quantified in terms of *interest rate*, money, probability, etc. After the transformation, made by the present value calculation, benefit—cost is still to some extent dependent upon discrete time, for example the time for the completion of a specific operation, but it is no longer a function of continuous time. Indeed, the calculation of present value is sometimes made using "piecewise" methods (more of this below) and the benefit—cost is then a function of discrete times, denoting conventional accounting periods, and these may be buried in the analysis.

Discounted cash-flow techniques are customarily employed in practical applications of benefit—cost analysis to calculate present value. That is to say, future rates-of-flow of cash are discounted at a rate of discount which is thought to represent depreciation of the real value of money with time to bring them to a value which represents their present worth. In the Public Service, a test rate of discount, usually recommended by the Treasury, is used. (In the U.K. the rate was 10% in 1971.) The private sector usually uses a discount rate which represents the opportunity-cost of capital to the organization. Investment decision rules, derived from classical economic theories of perfect competition, are usually not favoured for the pragmatic application of quantified analysis because imperfect competition in the real world ensures that there is a divergence between market prices and social benefits and costs.

Rules for the selection of discount rates representative of social time preference rates [1] for use in the benefit—cost analysis of public projects have never been satisfactorily resolved. In most cases it is customary to select a discount rate pragmatically and use it with the crude assumption that it is not invariant with time. In benefit—cost calculations, probability theory is sometimes used to predict the timing of discontinuities in cash-flow and is almost always used to obtain the *expected* level of rate-of-flow of benefit and cost. Hardly ever is probability theory used to allow for the possibility that discount rates may vary over time.

In both public and private benefit—cost analysis, some partial compensation for the last omission is made during the sensitivity analysis. Then interest, or discount, rate is occasionally allowed to vary but nearly always on the assumption that it varies uniformly over all time.

Discounting procedure is for some strange reason [2] usually recommended in discontinuous, "piecewise", form. Most organizations have copious arithmetic tables available which give binominal terms of (1+discount rate) raised to some negative power of the year at which the discount is to be made. With the aid of these, or their computerized analogue, analysts calculate the present value of benefits (not, at this stage, allowing for probability) from

$$\text{Present value} = \sum_{q=1}^{q=N} \pounds_q \, (1+r)^{-q}$$

Which is obviously compound interest reversed with r equal to the discount rate ($r = 0.1$ for 10%), N equal to the number of years from now in which the last significant cash flow occurs, and $\pounds_q$ represents the cash flow in the qth year. The negative of the same

---

[1] Social time preference rate would measure the willingness of the community to pay now, rather than later, for benefits sought.

[2] The reason might be that, a priori, the arithmetic of discontinuous discounting is more palatable to amathematical students. However, the cumbrous result of discontinuous discounting causes, in my opinion, such arithmetic and mathematical confusion at the stage of sensitivity analysis that the student would be more comfortable, in the long-run, if he learnt continuous methods at the onset.

128

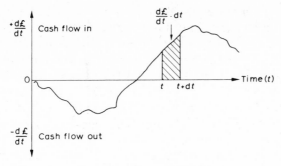

Fig. 1. A continuous cash-flow curve.

formula is used for the calculation of present value of cost and then $£_q$ represents cost accrued in the qth year.

Real cash flows are, of course, continuous and not composed of discrete flows at intervals of account (say every year as above). They may be represented as shown in Fig. 1. If discount is added continuously, the discounted value of $£_q$ obtained at the end of the qth year would be

$$£_q \left\{ \text{limit as } m \to \infty \text{ of } \left(1 + \frac{r}{m} \right)^{-qm} \right\}$$

which very conveniently is

$$£_q \, e^{-qr}$$

where e is the exponential constant, 2.718. The cash flow during the interval t to $(t+dt)$, the shaded area in Fig. 1, is $(d£/dt)dt$ and the summation of all possible shaded areas under the cash flow curve yields an integral for the present value of the cash flow

$$\int_0^\infty \frac{d£}{dt} e^{-rt} \, dt$$

where $d£/dt$ is the rate of flow of cash with time. Limits of the above integral need not be 0 and ∞ because real cash flows rarely start at zero and extend to infinite time. The last formula is the same as that for the Laplace Transform (see Jaeger (1951) for an explanation of the technique of Laplace Transformation which is much used in science and engineering and is usually integrated very

129

neatly for rudimentary functions of cash flow (d£/dt) usually assumed by benefit—cost analysts). When the function is non-analytic, i.e. cannot be integrated explicitly, it is still more amenable to numerical solution than the formula for discontinuous discounting would be applied to the same cash flow curve. Nascent analysts, who are not adept at elementary integration, should study Appendix 1 which contains present values of cash flows that commonly occur and a description of a numerical method[1] for the solution of non-analytic forms of cash flow.

However, the real significance of the continuous, or Laplace, method of discounting is that there are a large number of Laplace transforms already published for physical scientists which can be useful to the financial analyst or assessor of technology. Where the scientist transforms a quantity from a function of time to a function of frequency (and vice versa) the analyst transforms functions of time to functions of discount rate, which are easily manipulated in sensitivity analysis. Furthermore, the ubiquitous cash flows that crop up have present values which are easily remembered. Confirmation of this last statement may be obtained by observing the regular occurrence of certain analytic forms of present value, which occur in examples throughout this book, and noting how they make some aspects of benefit—cost analysis tractable.

It is also worth remarking that those who fear to challenge the analytic prowess of some assessors of technology can safely assume they are in a tortoise-versus-tortoise situation when the analysts are using discontinuous methods of discounting and not, as is too often assumed, in a tortoise-versus-hare learning mode.

For textual simplification a mathematical shorthand will be used throughout this chapter to denote present values. Discounted benefit, i.e. present value of benefit, will be written as

$$\text{Discounted benefit} = \overline{B(r)} = \int_0^\infty \frac{\mathrm{d}B(t)\,\mathrm{e}^{-rt}}{\mathrm{d}t}\,\mathrm{d}t$$

A benefit—cost quotient, i.e. the total discounted benefit divided by total discounted cost, will then be written as

---

[1] I claim that the numerical solution of present value problems using continuous methods of discounting is easier to apply than the discontinuous method. It is certainly true that in the former case only a very concise table of exponentials is required as opposed to the binomial tables used in the latter method.

$$\frac{\text{Present value of benefit}}{\text{Present value of cost}} = \frac{\overline{B(r)}}{\overline{C(r)}}$$

Using the same notation, total benefit net of total costs would be $\overline{B(r)} - \overline{C(r)}$. Note that unless it is explicitly stated, the text is always referring to present values of quantities.

Criteria

In some organizations, the absolute value of the benefit—cost quotient is taken to be indicative of the potential success, or otherwise, of a technological programme, for example, the PAU's statement quoted in Chapter 4 that "it has been found that the expected net returns (i.e. net of a further investment needed to exploit the research) on R & D investment are usually in excess of three times and frequently over five times the nominal R & D resource cost, including overheads, a finding which conforms with the estimates of several major industrial companies". The usefulness of examining the benefit—cost quotient for portent of success is that the quotient explicitly contains the discounted costs and these can be checked to see whether they satisfy a necessary constraint of being within the purse of the organization and commensurate with other portfolio investments.

Nevertheless, other criteria associated with benefit—cost calculation are used in "go", "no-go", decision making. Some organizations prefer to use the *internal rate of return* rather than the quotient in their decisions. The internal rate of return is simply the value of the discount rate which makes the benefit—cost quotient equal to unity or the present value of benefit net of cost equal to zero. Projects with internal rates of return greater than a minimal requirement and able to satisfy other requirements peculiar to the organization are then acceptable. Projects with smaller rates of return than the required minimum are not considered viable.

The arithmetic of calculating an internal rate of return from cash flows can be tedious especially when the discontinuous method of discounting to obtain present values is used. However, because the method has some supporters and is still in use it is necessary to give a short demonstration of the method.

An example of the calculation of internal rate of return

This example will be concerned exclusively with continuous methods of discounting for ease of exposition. Consider a cash flow which starts at time zero with a constant rate of spend equal to £A_c per annum and at time T becomes positive at a level of £A_b per annum. If the benefit is considered to flow for an indefinite time the present value will be given by case (10) in Table 1 of Appendix 1, with $T_1 = T$ and $T_2 = \infty$, i.e.

$$\text{Present value} = \left\{ \frac{£A_b}{r} e^{-rT} - \frac{£A_c}{r} (1 - e^{-rt}) \right\}$$

The value of the discount rate which makes the present value equal to zero is the internal rate of return

$$\text{Internal rate of return} = \frac{1}{T} \ln \left(1 + \frac{A_b}{A_c} \right)$$

which very conveniently is in a simple analytic form.

If the organization only authorizes projects yielding (say) a nominal internal rate of return of 20% before company tax, a project with $A_b = £\,25{,}000$ p.a. and $A_c = £\,50{,}000$ p.a. would have acceptance if the "breakthrough" time, T is equal to 2 years because

$$\frac{1}{T} \ln \left(1 + \frac{A_b}{A_c} \right) = \frac{1}{2} \ln(1.5) = 0.203$$

which is greater than 0.2. But if the breakthrough time, T, increases beyond

$$\frac{1}{0.2} \ln(1.5) = \frac{0.406}{0.2} = 2.03 \text{ years}$$

the project would not be acceptable.

Pay-back criteria

The last example enables an easy introduction to be made to another favoured method of evaluating projects, i.e. the pay-back method in which the criterion is the time taken to return the initial investment through the cash income. In the last example, if the pay-back time is designated T_p (and we still assume that the flow of benefit is almost indefinite) then the present worth of the initial investment will have been recovered when

$$\frac{A_c}{r}(1-e^{-Tr}) = \frac{A_b}{r}(e^{-Tr} - e^{-T_{p}r})$$

that is when the pay-back time T_p is given by

$$T_p = \frac{1}{r} \ln \left\{ 1 \Big/ \left[e^{-rT} \frac{1+A_c}{A_b} - \frac{A_c}{A_b} \right] \right\}$$

Thus if the discount rate $r = 0.1$ (10%), $T = 2$ years, and $A_c/A_b = 2$, we have

$$T_p = \frac{1}{0.1} \ln 1/(e^{-0.2} - 2) = 7.8 \text{ years}$$

And an organization with a pay-back time requirement of, say, seven years would reject the project. Projects with quicker pay-back are obviously preferred but it will be noted that in the last example the absolute magnitudes of the rate of flow of benefit and cost only entered into the calculation as a ratio. The quotient of benefit—cost is, generally speaking, preferred to pay-back or internal-rate-of-return as a criterion because it does show more explicitly the relative profitability of different projects with the costs explicitly displayed.

There is really no point in discussing the relative merits of present value of benefit net of cost versus the ratio of benefit—cost because, in the context of this chapter, the ingredients of both must be calculated separately before the final elementary steps of subtraction and division, respectively. Ranking projects in terms of net present value or benefit—cost ratio will generally give different preferences. For example

$B = 10$, $C = \frac{1}{2}$, therefore $(B-C) = 9\frac{1}{2}$ and $B/C = 20$ whereas
$B = 100$, $C = 50$, therefore $(B-C) = 50$ and $B/C = 2$.

But such anomalies are easily resolved when other constraints like availability of capital and effort and alternative opportunities are considered. The following examples should make this clear.

However, it is useful for the amathematical reader to note that the calculation of internal rate of return, or pay-back, can give some very anomalous results. Calculations can yield more than one positive rate of return, or pay-back time and even negative or purely imaginary, numbers. A discussion of the reason for such anomalies is not too mathematically involved but is tedious and will not be presented here. If any uncertainty arises during calculation of internal rates of return or pay-back, a mathematician should be consulted immediately.

Probability of Achieving Present Value

"If a man will begin with certainities,
he shall end in doubts; but
if he will be content to begin
with doubts, he shall end in certainities."
Francis Bacon, Valerius Terminus of the Interpretation of Nature

Discrete Probabilities

Technological projects are normally undertaken once and when they are unique, the probability of success in achieving benefit or constraining cost to a budget figure cannot be obtained from a large number of repeated trials. Classical methods of determining probabilities are very rarely available to the assessor of technology.

The so-called "probabilities" of benefit—cost analysis are, almost invariably, obtained from subjective estimates of an event occurring and these *may* be based upon past experience of similar projects. In the face of uncertainty, the analyst has only recourse to such estimates, although he may be fortunate in being able to calculate from mathematical first principles one or two of the probabilities needed. Customarily, the analyst treats the mixture of many subjective estimates and a few "real" probabilities as if they all obeyed the laws for the manipulation of "real" probabilities.

In this section, analytic procedures which allow for uncertainty will be illustrated with the aid of an idealized project consisting of a simple sequence of operations, each with a discrete probability of succes or failure (see Fig. 2, obtained from Alan Pearson, Manchester Business School).

Figure 2 indicates that only two possible outcomes are considered for each stage of the project, i.e. there is either success or fail-

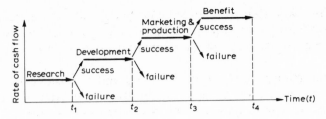

Fig. 2. Chronological flow-diagram of a simple project.

ure and no intermediate degrees of outcome are allowed. Probabilities of success are denoted by the p's and, since it is customary for the probabilities of mutually exclusive events to add to give unity, the probabilities of failure will obviously be given by $(1-p)$.

If the costs of the various processes, in chronological order, are $R(t)$, $D(t)$, $M(t)$, and the benefit is $B(t)$, the expectance of discounted net benefit will be

$$p_M p_D p_R \overline{B(r)} - (p_R p_D \overline{M(r)} + p_R \overline{D(r)} + \overline{R(r)})$$

where the (t)'s and the (r)'s denote whether the quantity depends upon time or discount rate. The formula contains products of "probabilities" because it is naturally assumed that success at the completion of a stage can only be achieved if the preceding stages have also been successful. Thus the probability of achieving benefit is $(p_R \cdot p_D \cdot p_M)$ and the probability of success at the end of the development stage is $(p_R \cdot p_D)$, whereas probability of success at the end of the research stage is simply (p_R).

No real project is as simple as the above example. Real projects may be composed of multiple connected, parallel, operations so that a chronological diagram looks like a spiders web. Some operations may even be straightforwardly sequential but because of idiosyncracies of the project's management structure, the above formula is likely to be incorrect. Real dilemmas arise, of course, when the project is intrinsically complex and also has rank bad management.

If the project is as defined in our simple example of Fig. 2 and the management is firm, so that the estimated probability of success of a particular stage refers to its completion (or cessation) in an *inelastic* period of time, calculation of present values will simply consist of integrations of cash flow over the estimated time period to completion.

As a heuristic example of two different means by which dis-

135

crete "probabilities" can enter benefit—cost analysis and the way the assumed efficiency of management can affect the outcome, consider a situation of simple sequential steps (as shown in Fig. 2) where

(1) the managers of the project stop and reconsider the project whenever a stage has exceeded the duration planned, and

(2) the managers have such loose control over the project that a stage tends to be repeated until success is achieved.

To make a comparative study it is useful to assume that both good and bad managers have started with the same estimates of the a priori probabilities of success. But in the case of bad management it is assumed that the probability of success, p, in each successive attempt at any stage remains constant and the time for completion of any stage is the estimated time multiplied by $(1/p)$.

Taking the estimated times shown in Fig. 2 and assuming that the time, (t_4), to the cessation of benefit is fixed (as is normally true when the product becomes obsolescent and there are competitors in the market) the following integrals for the present values of expenditure and benefit are obtained.

	Good management	Bad management
$\overline{R(r)}$	$\int_0^{t_1} \dfrac{dR(t)}{dt}\, e^{-rt}\, dt$	$\int_0^{t_1/p_R} \dfrac{dR(t)}{dt}\, e^{-rt}\, dt$
$\overline{D(r)}$	$\int_{t_1}^{t_2} \dfrac{dD(t)}{dt}\, e^{-rt}\, dt$	$\int_{t_1/p_R}^{[t_1/p_R + (t_2-t_1)/p_D]} \dfrac{dD(t)}{dt}\, e^{-rt}\, dt$
$\overline{M(r)}$	$\int_{t_2}^{t_3} \dfrac{dM(t)}{dt}\, e^{-rt}\, dt$	$\int_{[t_1/p_R + (t_2-t_1)/p_D]}^{[t_1/p_R + (t_2-t_1)/p_D + (t_3-t_2)/p_M]} \dfrac{dM(t)}{dt}\, e^{-rt}\, dt$
$\overline{B(r)}$	$\int_{t_3}^{t_4 (\text{obsolescence})} \dfrac{dB(t)}{dt}\, e^{-rt}\, dt$	$\int_{[t_1/p_R + (t_2-t_1)/p_D + (t_3-t_2)/p_M]}^{t_4 (\text{obsolescence})} \dfrac{dB(t)}{dt}\, e^{-rt}\, dt$

136

The excessive stretching of project time indicated by the right-hand column of integrals might be acceptable to the reader if he considers the way costs escalate in real technological projects. Mansfield *et al.* (1972) point out that, where a "large" technological advance is sought from a technical programme, the ratio actual cost/estimated cost can be as high as 4.2 and where a "small" technological advance is involved the average ratio is about 1.3. When Mansfield and his co-workers analysed forty nine projects carried out by a pharmaceutical company, they managed to obtain regression formulas for cost and time "overruns". Dependence of "overruns" on technological advance, derived from average respondent's estimates of difficulty rated on a five point scale, was found to be

$$\frac{\text{actual cost}}{\text{estimated cost}} = (\text{technological advance})^{0.51}.(\text{other variables})$$

$$\frac{\text{actual time}}{\text{estimated time}} = (\text{technological advance})^{0.21}.(\text{other variables})$$

Very roughly, the technological advance score over a single phase of a programme may be taken as inversely proportional to the probability of success. But as will be seen below from the development of the above example, technological advance (or probability of success in the technical parts of a programme) enter into benefit—cost calculations in complex ways. Nevertheless, Mansfield's statistical result gives some substance to the thought that bad project management, as depicted by the above expressions for present value, may be more prevalent than is commonly thought. Calculation of the expectance of discounted net benefit in the two cases will not be the same. Bad management will be expected to achieve

$$\overline{B(r)} - [\overline{M(r)} + \overline{D(r)} + \overline{R(r)}]$$

instead of the expected value given above which has included probabilities. Note that in the case of bad management the probabilities are buried in the integrals of present value and the prolongation of project phases, before benefit, has reduced (indeed it may negate)[1] benefit flow.

---

[1] If $t_1/p_R + (t_2 - t_1)/p_D + (t_3 - t_2)/p_M$ becomes greater than t_4 the present value of benefit becomes zero (see the limits of the bottom right integral above).

Numerical Example

To put the above algebraic expressions into quantified perspective (which is what benefit—cost analysis is all about) consider a simple sequential project with
rates of cash flow constant [1] and given by

$$\frac{dR(t)}{dt} = £1 \text{ million p.a.} \qquad \frac{dD(t)}{dt} = £\,10 \text{ million p.a.}$$

$$\frac{dM(t)}{dt} = £\,100 \text{ million p.a.} \qquad \frac{dB(t)}{dt} = £\,1000 \text{ million p.a.}$$

project timing of $t_1 = 5$ years, $t_2 = 10$ years, $t_3 = 15$ years and the time estimated for the obsolescence of the planned project (or service) as $t_4 = 25$ years, and
probabilities of success $p_R = 0.5$, $p_D = 0.75$, and $p_M = 0.9$.

(1) Calculation of Present Values Assuming Good Management.
With the assumption that the management is firm and will not allow habitual repetition of project stages the appropriate discounted values (see Appendix 1, Table 1), for a discount rate of 10%, are

$$\overline{R(r)} = \frac{1.0}{0.1}\,[1 - \exp(-0.1 \times 5)] \qquad = £3.93 \text{ million}$$

$$\overline{D(r)} = \frac{10}{0.1}\,[\exp(-0.5) - \exp(-1.0)] \quad = £23.9 \text{ million}$$

$$\overline{M(r)} = \frac{100}{0.1}\,[\exp(-1.0) - \exp(-1.5)] \quad = £145 \text{ million}$$

$$\overline{B(r)} = \frac{1000}{0.1}\,[\exp(-1.5) - \exp(-2.5)] = £1410 \text{ million}$$

Values like $\exp(-0.5)$ are easily obtained from tables of exponential functions such as those shown in Abramowitz and Stegun (1965), p. 116.

Therefore the expected present value of benefit, net of cost, is

$$p_M p_D p_R \,\overline{B(r)} - [p_D p_R \,\overline{M(r)} + p_R \,\overline{D(r)} + \overline{R(r)}] = 406 \text{ million}$$

---

[1] The assumption of constant rate of flow of expenditure or benefits is extremely naive. Some of the trapezoidal, exponential, or sine-wave shapes (given in Appendix 1) would be more appropriate. But I do not wish to clutter the numerical demonstration with too much algebra.

Using a different arithmetic combination of benefit and cost, the quotient of expected benefit to expected cost is

$$\frac{E(\overline{B(r)})}{E(\overline{C(r)})} = \frac{p_M p_D p_R \overline{B(r)}}{(p_R p_D \overline{M(r)} + p_R \overline{D(r)} + \overline{R(r)})} = \frac{476}{70} = 6.8$$

(2) Calculation of Present Values Assuming Bad Management. With the assumption that the management is poor and will allow habitual repetition of project stages until an "apparent" success has been achieved, the appropriate discounted values are

$$\overline{R(r)} = \frac{1.0}{0.1} [1 - \exp(-0.1 \times 5.0/0.5)] = \text{£}6.3 \text{ million}$$

$$\overline{D(r)} = \frac{10}{0.1} [\exp(-1.0) - \exp(-0.1(5.0/0.5 + 5.0/0.75)]$$

$$= \text{£}17.6 \text{ million}$$

$$\overline{M(r)} = \frac{100}{0.1} [\exp(-1.66) - \exp(-0.1(16.6 + 5.0/0.9))]$$

$$= \text{£}81 \text{ million}$$

$$\overline{B(r)} = \frac{1000}{0.1} [\exp(-2.21) - \exp(-0.1 \times 25)] = \text{£}280 \text{ million}$$

Therefore the expected benefit, net of cost, is

$$\overline{B(r)} - [\overline{M(r)} + \overline{D(t)} + \overline{R(r)}] = 280 - 105 = \text{£}175 \text{ million}$$

Using the quotient of

$$\frac{\text{Expected benefit}}{\text{Expected cost}} = \frac{\overline{B(r)}}{\overline{M(r)} + \overline{D(r)} + \overline{R(r)}}$$

$$\frac{E(\overline{B(r)})}{E(\overline{C(r)})} = \frac{280}{105} = 2.7$$

(3) Comment on the Example. Placed in juxtaposition, the above arithmetical results are as follows.

	Good management	Bad management
Net present value	£406 million	£175 million
Expected benefit / Expected cost	6.8	2.7

Although the example has been carried out "tongue-in-cheek", these large discrepancies are alarming! The assumption that the obsolescence time, when benefit terminates, is constant and not pushed farther into the future is, obviously, a major reason for the differences. However, even if it is assumed that benefit is only delayed and not truncated by bad management, there will still be discrepancies although, naturally, they will be smaller. In real life the penalty for procrastination, or forced delay, will be twofold, a truncation of time over which it is possible to obtain benefit and a reduction in the level of benefit. The example is, therefore, too favourable to bad management in keeping the rate of flow of benefit constant. Analysts in the same situation would try to introduce more allowance for likely success and lead time success, of *known* competitors[1].

Probability techniques are commonly used when planning the project's logistics and calculating expectancies (as in the first part of the example) but it is uncommon for them to be used to throw light on the consequences of delay. Because analysts are not usually the managers of the projects they analyse, such restricted use of probability theory is likely to give erroneous results. Furthermore, note that if the management's quality is uncertain, the use of any kind of probability calculation only affirms that the result must lie between the extremes of good and bad shown above. The former calculation demonstrated is invariably optimistic and yet it is the kind most often exhibited in benefit—cost calculations. If readers will scan published benefit—cost analyses they will note that discrete probabilities are ubiquitously used to adjust expected levels, and not durations, of cash flows. Perhaps this is one reason why "mammoth" technological projects escalate far beyond expectations in cost and time?

---

[1] Without reasonable support from colleagues carrying out industrial intelligence (see Chapter 9) on the strength of competition, analysts are in danger of feeding unrealistic assumptions into their work.

A rudimentary demonstration of sensitivity analysis may be made with the use of the above example. Suppose that the estimate of probability of success in marketing and production is in dispute and it is known that the project will have good management. If all the other probabilities are agreed, then a sensible step is to repeat the arithmetic of part (*1*) of the example with different values of p_M to see how sensitive the result of the calculation is to the choice of p_M.

If the criterion of project assessment is the expected value of net present benefit, then simple arithmetic shows that the difference in outcome for two different values of p_M is $p_R.p_D.B(r)$ (difference in the values of p_M). Thus a project with assured success in marketing and production (with $p_M = 1$) will yield the following advantage, in expected net-benefit, over a project with an estimated $p_M = 0.9$

$530(1.0 - 0.9) = £53$ million

Whereas a project with $p_M = 0.6$ will yield the following disadvantage

$530(0.6 - 0.9) = -£159$ million

where the minus sign signifies a loss. Sensitivity analysis does not usually yield such easy answers and although arithmetic will always give a result at the cost of excessive labour, the analyst usually relies upon the mathematical process of differentiation (see the following section of this chapter on sensitivity analysis).

Continuous Probability Distributions

Although the predominant part of benefit—cost analysis is usually carried out with discrete subjective estimates of an event occurring treated as discrete probabilities (i.e. a consensus that an event has a one in ten chance of occurring is taken as an a priori probability of 0.1), more sophisticated treatments of probability analysis do occur. For instance some analysts, with a background in the theory of statistics, favour the use of probability distributions in continuous form or, more exactly, the use of probability density (or sampling) functions (PDFs).

This later technique usually begins with the selection of appo-

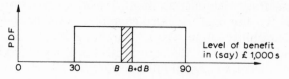

Fig. 3. A sampling function.

site theoretical pdfs from statistical theory or with the construction of PDFs from a sufficient number of respondent's estimates of an event occurring. A typical example would be that shown in Fig. 3. Here the probability of the benefit lying between B and $B + dB$ is the area under the PDF curve between these limits and the area under the whole curve is unity, because the benefit must lie between 30 and 90.

Examples of this rectilinear type abound and they are obviously poor representations of real situations because one would not expect the PDF to cut off so sharply at its extremities. When they occur in practice it is because there is such lack of perception that it is assumed by the analysts, after evaluation of respondent's estimates of benefit, that it is equally probable that the benefit will lie anywhere between the minimum and maximum estimates of 30 and 90 thousand pounds per annum. (Rectilinear PDFs occur frequently when Bayesian Statistics are used in the presence of acute uncertainty but it is not the author's intention to discuss these in this book.)

Analytically speaking, it is not too difficult to handle a PDF with the help of a competent statistician. However, when calculating benefit—cost quotients or other arithmetic combinations of benefit and cost, there are many input parameters in the analysis, each with an associated PDF, and these *must* be combined into some overall PDF or sampling distribution. Combinatorial analysis, as such calculation of sampling distribution of some function of independent variates is popularly called, can be an extremely messy business but fortunately there are well-established numerical techniques which may be successfully applied to the task with the aid of a digital computer. One simply tells the computer what the individual PDFs and the rules for the combinations of the independent variates are and the machine simulates the overall sampling distribution and provides graphical or numerical output. Sometimes the task, if not beyond the capability of the computer, is a trial for the computer programmer because of the inherent discontinuities in the PDFs of the individual variates (like the

142

sharp cuts off in the above diagram). Notwithstanding these difficulties, the programmer usually succeeds in his task and the arbiter-of-decision is presented with a sampling distribution of theoretically comforting mien.

An arbiter, presented with an overall sampling distribution of benefit—cost quotient, say, should ideally be armed with some means of probing this precipitation of the analysis. For this reason, and to encourage better perception of the process, the following graphical exposition of the differently shaped sampling distributions for the quotient benefit—cost in a situation of minimal complexity when both benefit and cost have rectilinear PDFs is made[1].

Example

The probability density functions of benefit and cost are assumed to be overlapping, as represented in Fig. 4. Heights of these

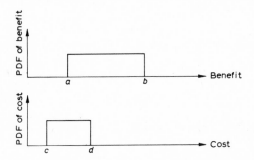

Fig. 4. Sampling functions of benefit and cost.

PDFs have not been given because they are derived from the normalising condition which is that the area under the PDF curves should be unity because it is assumed that the probability of benefit (say) occurring between the limits a and b is 1, i.e. the benefit *must* lie somewhere between these limits. Therefore the heights of the PDFs are $k_b = (b-a)^{-1}$ and $k_c = (d-c)^{-1}$ for benefit and cost respectively.

---

[1] Readers interested in the analytical derivation, made in this case by the author and his PAU colleague Keith Taylor, of the sampling distributions presented should consult Appendix 2 where they will find expressions of the means and variances of the different distributions.

143

If, for convenience, we define the benefit—cost quotient as u we can obtain the analytic form for the sampling distribution of benefit—cost by using the theory of statistics, in particular that part which defines the sampling distribution of the ratio of two independent variates (see Kendall and Stuart, 1963). A slight difficulty arises in the application of the theory, however, because an integral needs to be solved for the sampling distribution of the quotient and the finite limits of the integral depend upon the relative magnitudes of the ratios given by the extremities of the PDFs of benefit and cost, i.e. upon the relative magnitudes of (b/d) and (a/c). (Mathematical enthusiasts see Appendix 2.) The three possible rankings of relative magnitude are (b/d) is larger than (a/c), (a/c) is larger than (b/d), and (a/c) is equal to (b/d). Each of these conditions leads to a differently shaped sampling distribution for the quotient of benefit—cost, u as shown in Fig. 5.

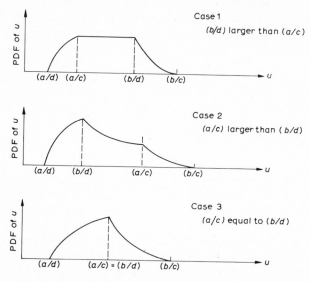

Fig. 5. Sampling functions of benefit/cost quotient.

An arbiter-of-decision supplied with the above sampling distributions could hardly be criticized if he assumed that the results obtained were from entirely different types of distribution of benefit and cost. He could not be expected to know that the different shapes were obtained from input distributions which all had the same general shape and are a consequence of different assumptions made about the cut off values of these distributions.

144

In order to make better use of his analytic support, the recipient of a sampling function should always try to discover how the sampling distribution was arrived at. He should frame probing questions about any singularities he perceives. Indeed, if the distribution has no singularities he should ask why not. When all else fails he must ask the analysts what the outcome would be of changing one of the input distributions. For example, if the analyst has used a smooth normal distribution, from statistical theory, the questioner might ask what the consequence would be of replacing the normal bell-shape with a rectangular distribution of the same half-amplitude width of the bell.

It is in the "twiddly-bits" of analysis that the arbiter has much to fear and the bad analyst much to hide. A good maxim is to challenge all precipitations of the analysis to obtain, at the least, a verbal understanding of the nature of the precipitate. Even a cursory understanding of the rudiments of analytic technique will enable this to be done and hand-waving obscurantists of theory can with effort be utilized as tutors and made into more efficient advisors by the amathematical decision maker who possesses a thick skin. This is one game that the discriminating decision maker cannot lose because his good analysts will be flattered by his attention and the mediocre analysts will improve if they suspect that they are under sensible surveillance [1].

Sensitivity Analysis

The reader should have gathered by now that benefit—cost analysis is not an exact science and is highly coloured by the foibles of the analyst. To emphasize this point and demonstrate more complicated sensitivity analysis in use in a situation where there is incomplete data, a benefit—cost model for prospecting and mining (undersea) made by Medford (1969) will be reiterated.

Example
Certain critical parameters control the benefit—cost ratio of

---

[1] This is not just avuncular advice. I have been very impressed with the contribution made by intelligent generalists to the analytic content of assessments. Conversely, I have been dismayed to note the effect of some partially numerate arbiters on analysis when, rather than admit unfamiliarity with an analytic technique, they condone its use without question.

a project and it is usually possible to assess objectively the economic viability without recourse to mathematical formulation when not more than five of these parameters are variables. When a larger number of parameters is involved it is better to build a mathematical model which describes, in simplified form, the crucial aspects of the problem.

The parameters which control the viability of off-shore mining operations are

A Area of sea bed prospected, in thousands of square miles.

X Attainable throughput of processed mineral, obtained from A, in thousands of tons per annum [1].

p Probability of achieving throughput X from area A.

M Market price of the mineral in thousands of pounds sterling per ton.

R Cost of recovering mineral from the sea bed, in thousands of pounds sterling per ton. (Derived from the cost of recovering gangue and ore and the average tenor.)

T Transport cost, in thousands of pounds sterling per ton of mineral.

L Cost of "mineral rights", in thousands of pounds sterling per ton of mineral.

r Discounting factor, i.e. r per cent/100.

The first step in the exploitation process is to submit the designated area A to saturation prospecting. Such a process is inevitably painstaking and will obviously be a major time-consuming factor in the exploitation of undersea mineral wealth. At the time this model was conceived, no accurate guides which would enable the rate of prospecting to be estimated were available to the modeller (a common state of affairs in the field of benefit—cost analysis). The best informed guess was that sea bed prospecting will not proceed faster than terrestrial prospecting in the Yukon (say 1,660 square miles per annum at an annual cost of $ 1 million).

After $t = 0.6A$ years, therefore, it is assumed that area A has been satisfactorily prospected and that the costs have been incurred at a rate of approximately £ $0.5S$ million per annum, where S is a multiplier which allows for the additional difficulties experienced by marine prospectors (see Appendix 1, Medford, 1969) and is currently equal to 10 for the U.K. continental shelf and somewhat smaller for calmer waters.

Each region within area A possessing geophysical portent will be

---

[1] For minerals such as gold and silver, more convenient units are oz. and lb.

submitted to drilling evaluation and the cost of this will be proportional to the expectation of throughput pX. Drilling evaluation costs are therefore put equal to £pXk million; where k is a factor (see Appendix 1, Medford, 1969) which allows for the type of drilling required (shallow or deep, etc.) and can also allow an allowance for technological improvements in drilling technique (e.g. turbine drills).

It should be noted that evaluation costs are proportional to the expectance of achieving an ultimate throughput of X of mineral, i.e. pX, and not to the size of the area prospected. (One only drills where it is highly probable that minerals will be found.)

A mining system will be required to work the deposits revealed by drilling and for the purpose of this analysis such systems have been divided into categories of use which are extensions of dredging technology, undersea shaft mining, and mining by drilling (i.e. oil, gas, sulphur, etc.).

On the assumption that the engineering of mining systems will be akin to shipbuilding and construction of chemical plant, the capital cost of the mining system is expected to be proportional to (expected throughput)$^{0.6}$, i.e.

$$C = C_s (pX/X_s)^{0.6}$$

where C and C_s are capital costs in millions of pounds for annual throughputs X and X_s respectively.

It is assumed, inaccurately, that the basic system cost will not vary significantly with environmental severity of the operating zone. However, the major cost induced by environmental severity is on the utilization rather than the capital costs and the adjustment for this is made below.

The production of the mining system will be highly dependent on outage, i.e. days on which production is low or zero, which will depend on both the mechanical reliability and the environmental conditions. These are considered in Medford (1969) and a factor q combining both these factors and relating to the environmental zone is there derived.

These considerations result in the following cash flows.

(1) The rate of flow of cost will be approximately constant over $0.6A$ years and equal to

$$£\ 0.5S + \frac{pXk}{0.6A} \text{ million per annum}$$

147

(2) At that time $t = 0.6A$ years $£C_s(pX/X_s)^{0.6}$ will be spent on mining systems.

(3) After $0.6A$ years benefits will flow at a rate of

$$£ \ q(pX)(M-(R+L+T)) \text{ million per annum}$$

Incorporating these cash-flows into a benefit—cost ratio gives after algebraic simplification, the following model for the economics of marine mining.

$$\frac{E(\overline{B(r)})}{E(\overline{C(r)})} = \frac{q(M-(R+L+T))}{r(C_s/pX)(pX/X_s)^{0.6} + ((0.5S/pX) + (1.67/A))(e^{+0.6A}-1)}$$

Figure 6 shows schematically the cash flows incorporated into this benefit—cost model for the ratio of net revenue to cost of exploitation; the transformation of these cash flows into present values may be followed with the aid of Appendix 1 of this chapter.

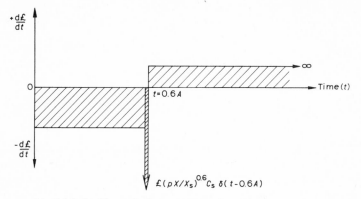

$\delta \ (t-0.6A)$ signifies a single payment at time $0.6A$ years

Fig. 6.

The model was checked against the known economics of existing off-shore operations for the extraction of sand/gravel, oil, and gas and was found to be sufficiently accurate as a base for sensitivity analysis. By taking statistics of cost of modern engineering equipment, costs of recovery and transportation together with the market price of the mineral to be extracted, it is possible to reduce the mathematical model to a function of area prospected, A, and expectance of throughput, pX. For example, for the mining of tin on the U.K. continental shelf the following statistics were inserted into the model.

148

$C_\mathrm{s} = 3.3$, $X_\mathrm{s} = 0.4$, $M = 1.5$, $R = 0.18$, $T = 0.02$, $L \to 0$, $k = 0.066$, $r = 0.08$, and $q = 0.6$

From the resulting arithmetic expression it was possible to examine the sensitivity of benefit—cost to changes in the expectance of throughput and area prospected. The graphical outcome of the sensitivity analysis is shown in Figs. 7 and 8. Obviously the maximum benefit—cost occurs for a given expectance of throughput when $A = 0$, i.e. when the location of tin-bearing strata is known and no prospecting is necessary. The curve for an expectance of throughput of 1,000 ton/year clearly denotes an unprofitable operation when the area prospected is greater than 400 square miles, and even if prospecting is unnecessary the maximum benefit—cost ratio is only 1.1, a commercially unattractive rate of return at a discount rate of 8%.

It needs to be stressed that the benefit—cost model described above was only concerned with commercial aspects of undersea mining. Even if the model included extra terms, in the benefit, for the saving of foreign currency on mineral importations to the

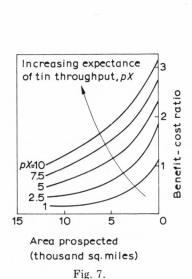

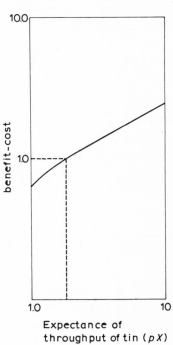

Fig. 7.

Fig. 8.

country and the provision of work for the unemployed, it would not be an adequate model in the collective social sense because it contains no terms for the possible destruction of amenity due to possible pollution or depredation of seascape/landscape. In examining a benefit—cost analysis, one should always ask, to whom do the stated benefits accrue and who meets the costs[1]?

Suppose that the analysis is commissioned by a government department. Then one would expect a model, similar to the above example, to be made as a first step in the analysis simply to put the assessors into a position where they could understand the mining entrepreneur's motives. Ideally, another model would then arise, out of the arbiter—analyst dialogue, which would include.

(1) the national benefits anticipated from the entrepreneur's efforts, such as increased employment, savings in foreign currency, and the possible inducement to secondary industries, and

(2) all the costs to the nation, including investment grants, tax incentives, opportunity costs of labour attracted to the undersea mining venture, *and* the social cost of reducing a collective amenity or increasing pollution, etc.

Responsible entrepreneurs would not, of course, fail to make their own calculations (if only notionally) of some of these latter factors because they need to know the forces opposing a venture before committing their backer's money; or more idealistically, they will be spurred by their own social consciences. At the present time there is neither widespread expertise, or facility for the construction of benefit—cost analyses which adequately allow for social factors, or a forum with audience and orators possessed of a sufficient level of comprehension of such analyses (even those rudimentary analyses which have already been conceived). *The bludgeon of rhetoric still triumphs in some quarters over the rapier of analysis and the public is the ultimate sufferer!* Occasionally the rapier, or something that looks remarkably like it, wins unfairly as

---

[1] Dr. Milton Katz, Director of International Legal Studies, Law School of Harvard University has commented succinctly on this question, See Katz (1970). For example, "I would like to emphasize one point about the assessors who are, in fact, the whole business community. Whether or not they are aware of the fact, the existing legal order infuses their calculations. It is the legal order that determines which of the anticipated costs and benefits are taken into account by the enterprise and which can be ignored .... The economic mode of analysis is an indispensable tool for technology assessment. But I want to emphasize that the economic analysis takes for granted a particular posture of the existing legal system". (See Chapter 2.)

the following quotation on the arithmetic ranking of subjective opinion (see Chapter 7) demonstrates.

> "This process leads to decision outcomes that have such clarity and force that they can withstand even the strongest questioning, as Congress was taught in the controversial TFX investigations. This situation illustrates the hazards produced when the lack of facts leads to overcompensation in elaborate techniques for handling non-facts."
>
> Edward B. Roberts

Sensitivity Analysis of Technology

Any benefit—cost model incorporating technology should give some indication of the likely impact of technology on the system modelled. From the first-order marine mining model discussed above, for example, it is possible to obtain an indication of the sensitivity of the benefit—cost ratio of marine mining operations to advances in technology. Clearly any technical improvements or innovations in instrumentation for prospecting, drilling evaluation or recovery equipment will reduce the cost of ore in the "bin". The question is how much?

No complete list of improvements exists, or is ever going to be available; moreover, such improvements as do exist are likely to advance the cause of both marine and terrestrial mining. It is possible, however, to point to embryonic advances in technology which should be nurtured and suggest that spending on new technology be judged from the probable increment in benefit—cost resulting from implementation of technological goals, assuming these have been achieved. Straightforward sensitivity analysis of the model shows, for example, that the ratio of increments in benefit—cost resulting from enhancements dk and dS in drilling and prospecting factors is

$$\frac{\Delta(B/C) \text{ due to improvements in drilling evaluation}}{\Delta(B/C) \text{ due to improvements in prospecting}} = 3.33\left(\frac{pX}{A}\right)\frac{dk}{dS}$$

Thus in assessing two competing projects, one designed to improve prospecting by dS and the other to improve the efficiency of drilling evaluation by dk, we would choose the former if 3.3 $(pX/A)(dk/dS)$ is less than one and the latter if it is greater than one.

Another ratio which is pertinent to decisions of this type is

$$\frac{\Delta(B/C) \text{ due to improvements in drilling evaluation}}{\Delta(B/C) \text{ due to improvements in recovery technology}}$$

$$= \frac{1.67[\exp(+0.6rA)-1]\,dk}{A(pX)^{0.4}\,X_s^{-0.6}\,dC_s}$$

This ratio is functionally dependent on two more variables, one of area prospected and the other of annual throughput of mineral.

Quantitatively, in the original analysis, the analysts were not able to define dk, dS, and dC_s for potential innovations which were known but they were able to list some of the likely innovations quantitatively under appropriate headings (see Medford, 1969). If the listed innovations are ever placed in a position of competing for funds to develop the technology, the above sensitivity analysis represents one way of comparing investment opportunities per capita of investment.

Apart from the details peculiar to the mining industry, the last example is easily understood in principle. Similar procedures of sensitivity analysis are usually carried out without much difficulty whatever commercial proposition is considered. When the proposition involves social benefits, however, these are not easily quantified yet there is possibly a greater need for some form of sensitivity analysis in such a case.

Examples involving quantification of social benefit which are instructive and believable are almost impossible to find. At this stage in the development of quantification of social benefit—cost no analyst would flaunt any of his conceptual models without sounding excessive notes of caution. The model presented below, *for demonstration purposes only*, makes no pretence at accuracy and lacks factors in the social goal such as satisfaction from the access to pollution-free environment. It should serve to show how very complex are the problems of modelling social benefit—cost relationships.

A Speculative Sensitivity Analysis of Social Benefit

Decision making in the field of education excites much controversy about social benefits and disbenefits. This example uses a very small part of the report NCET (1969) made by a British study team which was set up in 1968 to study the national impact of computer-aided instruction systems (CAI or educational technology which uses digital computers).

152

The team consisted of three members of Programmes Analysis Unit, three representatives of Education, and one from Industrial Training. Their primary task was to examine the feasibility of a possible large scale national programme in computer-based learning systems.

Early in the investigation it became clear that computer-based learning systems in the U.K. and elsewhere had not provided sufficient research evidence of their effectiveness and the team was necessarily concerned with judgements rather than proof, with what "might" be rather than what "will" be. The team met this problem of assessment by two approaches. First, it considered the potential use by a subjective projection, i.e. a scenario, of what might be developed both technologically and educationally. Secondly, it examined the consequences of problems and decisions which will arise from the development in the U.K. and abroad of the use of computers in education. From these approaches the team derived the nature and hence the probable cost of development both to meet particular educational aims and for the minimum defensive strategy.

An interesting off-shoot of this study was that the team were specifically asked to indicate acceptable techniques of benefit—cost and cost—effectiveness analysis which could be used for assessing the value of projects in computer-aided education. The imposition of this task upon the team caused vigorous discussion and there was a dichotomy of opinion about whether education was of more value to the society as a means of satisfying working needs, or of producing good results in the "affective domain" of behaviour. These opposing views were only reconciled within the team after an iconic[1] model was constructed which showed the part education plays, or should play, in our society. Figure 9 shows the model constructed. It can be simply defined as a multiple feedback system with only one non-closed output which represents the ultimate goal of leading the "good life" (defined according to the ethics of the era).

---

[1] From the Greek "eikon". An iconic model is a graphical representation of the real world, usually recognised by its pictorial appearance of lines, with arrows, connecting labelled spaces. Such an iconic model, without quantitative development, can be criticized as being as useful a guide to analysis as a Jackson Pollack painting. But even so, the analyst should not be deterred from making such models if these stem from a genuine desire to find out what goes on "inside the boxes" and not a desire to bury the problem in an icon-like pretty picture.

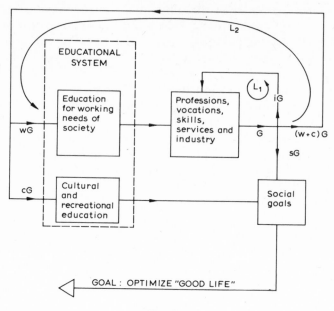

Fig. 9.

Only two of the loops, representing feedback in Fig. 9, are positive in the sense that they can sustain themselves. These have been labelled L_1 and L_2.

L_1 is the primary loop which sustains our society in the absence of harmful exogenous effects and produces Gross National Product (denoted by G in the model).

L_2 is the "control" loop which allows our society to adjust and create new thought and technology and protect itself from harmful exogenous effects (for example, superior technologies of competing societies).

Although the whole iconic model is bestrewn with value judgements, the most important one, in my opinion, is that L_2 has such importance that without it the judgement is that there can be no successful maintenance of the "good life" since we need both intellectual and financial inputs to achieve this social goal.

In trying to assess the benefits likely to be obtained from changes in the education system, or indeed any other system, consideration should be given to the possibility of

154

(*1*) providing the same service at different cost[1],

(*2*) being able to make more effective use of the resources available (e.g. buildings, equipment, teachers), and

(*3*) producing a better product.

In the field of education it is particularly difficult to specify this "better product". How can the required qualities in a "economic man" be specified educationally? Does an "economic man" necessarily mean an "ideal man"? Does the method and content of the educational system affect the product? Until these problems have been resolved it is impossible to calculate the potential benefits caused by a technological change in the system with any accuracy. An exact calculation of the expected benefit of CAI on "National Benefit" or, to be more explicit, the quantification of the likely contribution which could be made to the Gross National Product or towards Social Benefit is, therefore, an unsurmountable task at the moment. Two cogent reasons, in themselves sufficient, are

(*1*) that estimates of potential educational efficiency of CAI will not be universally agreed unless more meaningful, controlled experimentation gives a reliable measure of efficacy and

(*2*) that the causal relationships relating changes in educational efficiency to economic and social benefits are not yet understood.

The first reason is well understood by educationalists, experimental psychologists, and the multi-disciplinary protagonists of CAI and requires no further explanation here.

The second reason deserves more careful examination because it predicates the need for economic research which could eventually lead to a rational basis for making decisions throughout the whole of our educational system.

A simple attempt to correlate growth of GNP at a given time with earlier expenditure on education would not give a sufficient answer even if we are able to believe that the growth of GNP represented true benefit to the nation. For example, a regression analysis could not possibly yield an acceptable result unless causal

---

[1] This was done in the original study NCET (1969) and is simply a pragmatic examination of whether the installation of a Computer-Aided Instruction system would give an expected change in annual expenditure plus an accompanying and quantifiable educational benefit which shows a positive total benefit.

An exposition of such pragmatism is not relevant to the purpose of the present discussion, but the sums did show that economic considerations favoured the introduction of CAI in the order: Colleges of Education; Further Education and Secondary Schools; and, finally, Primary Schools.

relationships were defined and all the inherent critical parameters are known. The search for these causal relationships constitutes a large portion of the necessary economic research which still needs to be refined.

Despite the above reservations the National study team had to at least try to discharge its remit and an attempt was made to insert some causality and quantification into the iconic diagram of Fig. 9. By initially confining interest to that part of the hypothetical system formed by the two positive feedback loops L_1 and L_2, it was possible to obtain some analytical comprehension of the hypothesis as follows. The communality between loops L_1 and L_2 is that they both pass through that part of the system which creates wealth (G). This is schematized in the figure by the box labelled skills, services, and industry. Loop L_1 feeds capital and loop L_2 feeds intellectually and physically skilled manpower into the wealth producing part of the system (middle box in Fig. 9).

Skilled manpower may be denoted by

$$M = zN \tag{1}$$

where N is the working population and z is the average working efficiency.

Capital feedback is defined as

$$C = iG \tag{2}$$

where i is the fraction of GNP feedback as fixed assets.

If it is assumed that GNP is a linear scalar function of skilled manpower and capital we can use the economist's Cobb-Douglas Function and then it follows that

$$G = AM^a C^{(1-a)} \tag{3}$$

where A and a are "constants" whose values depend upon the competitive interaction within and without our society.

The fraction of the GNP fed back into education for the *working*[1] *needs of society* is designated w and thus the money per worker, spread over the whole population, devoted to non-cultural

---

[1] To aid the semantics, scientific and technological skills are included as a necessary part of the working needs of society and, for convenience only, labelled non-cultural.

156

education is wG/N. Furthermore, *if* education produces an average functional efficiency h for each pound spent, the average working efficiency is

$$z = h(wG/N) \tag{4}$$

Substitution of the values of M and C into the formula for the production of GNP yields a modified Cobb-Douglas Production Function.

$$G = Ah^a \, (wG_0)^a \, (iG_0)^{(1-a)} \tag{5}$$

where G_0 represents the GNP at the time of feedback; this is to say that it is assumed that feedback is quantized and the G_0 represents the GNP at some previous time, one average cycle of labour and capital before it is required to calculate G. (This is, of course, a gross simplification but a more careful analysis allowing for a continuous process of feedback would yield a differential equation and not provide particularly more insight into the system.) If we denote the increment of GNP during the "cycle" as

$$P = (G/G_0) \tag{6}$$

eqn. (5) becomes

$$P = Ah^a w^a i^{(1-a)} \tag{7}$$

In the iconic diagram the fraction of the GNP fed back directly into communal social welfare is defined as s and the fraction fed into cultural and recreational education is c, therefore

$$1 = w + i + c + s \tag{8}$$

and assuming that the social goal is proportional to $(c + s)$ we find

$$\text{Social goal} = [1 - (w + i)] \, G \tag{9}$$

Note that the contribution to the social goal from cultural education has not been weighted with the efficiency of education. It could be, but since people participate in this type of activity under their own volition it may not be appropriate to allow for an enhanced efficiency in the educational process that depends essentially upon a certain degree of regimentation. (Any reader who disagrees with this statement is welcome to insert the functional efficiency of education directly into the definition of the social goal. He will not find the algebra difficult.)

Elementary calculus shows that this goal is optimized if:

$$w_{\text{opt.}} = \frac{P}{Ah}\, a \left(\frac{a}{1-a}\right)^{(1-a)} \tag{10}$$

and

$$i_{\text{opt.}} = \frac{P}{Ah}\, a \left(\frac{1-a}{a}\right)^{a} \tag{11}$$

The maximum expenditure possible on communal cultural welfare and cultural and creational education is then

$$G_0(c+s) = PG_0\left(1 - \frac{w_{\text{opt.}}}{a}\right) \tag{12}$$

or

$$\text{maximum expenditure on social goal} = PG_0\left\{1 - \frac{P}{aAh}\, a\left(\frac{a}{1-a}\right)^{(1-a)}\right\} \tag{13}$$

Discussion

The eqns. (10)—(13) inclusive give some qualitative and quantitative appreciation of the importance of education in our society. Reference to eqn. (5) shows quite clearly the original value judgement fed into the hypothesis but, given that the relationships used are not too far wrong, it goes somewhat further and gives some quantified "feel" for the situation where lack of success in external competition leads to a reduction in the values of A and a (this is highly likely although I do not know of any statistical evidence that this is so). If A and a are changed in such a way as to depress the GNP, compensator changes must be made in the efficiency of education, h, or the amount of GNP fed back into education and/or capital investment must be increased. Similarly, the maximum allowable expenditure on social goals must be reduced in the face of too strong an external competitor (see eqn. (13) which becomes a monotonically decreasing function with reductions in A and a).

An interesting development of the equations for optimum social benefit is that the enhancement of social benefit for an increment of educational efficiency is

$$\frac{\text{Enhancement of social benefit}}{G_0} =$$

$$\frac{P^2}{aAh^{(a+1)}}\, a \left(\frac{a}{1-a}\right)^{(1-a)} \times \text{enhancement of educational efficiency}$$

The benefit—cost ratio of any project aimed at improving the functional efficiency of education would then be derived as follows.

$$\frac{\overline{B(r)}}{\overline{C(r)}} = \frac{\text{Present value of continuing enhancement and social benefit}}{\text{Discounted costs of research, development and installation, and annual costs of implementation}}$$

In the case under discussion this reduces to

$$\frac{\overline{B(r)}}{\overline{C(r)}} = \frac{\dfrac{G_0}{r}\,\dfrac{P^2}{Ah^{(a+1)}}\left(\dfrac{a}{1-a}\right)^{(1-a)}}{[\overline{R(r)} + \overline{D(r)} + (1/r) \times \text{annual costs of implementation}]}$$

where $R(r)$ and $D(r)$ and r are the discounted research and capital costs and discount rate respectively.

To make the formula for benefit—cost and its use a little more explicit, consider the case where the society knows how to dispose its resources optimally in order to optimize social benefit[1]. Then the only *new* educational technology projects which can be countenanced are those which satisfy the following inequality, derived from a rearrangement of the above equation and the constraint that the quotient of benefit—cost should at least be greater than unity.

---

[1] An analyst working for political masters can only assume that this is so if he wants to work in his post long enough to help further decisions.

Research costs + capital costs of Development and Installation must be less than

$$(1/r) \left[G_0 \; \frac{P^2}{Ah^a} \left(\frac{a}{1-a}\right)^{(1-a)} \frac{\mathrm{d}h}{h} - \text{annual costs of implementation} \right]$$

The use of this inequality may be demonstrated as follows.

Suppose that the enhancement of GNP is so small that P^2 is very nearly equal to unity and that in the pre-technological innovation stage the constants of the Cobb-Douglas function would have given that part of the first term in the bracket which is equal to w_{opt} (see eqn. (10)) the value of 1/20 (i.e. 5% on education for working needs). Furthermore, let us assume that the vendors and educational assessors of a new computer-aided instruction equipment are agreed that the new technology is sound from the viewpoint of engineering reliability and educational efficacy. Although the educators are not able to measure the absolute efficiency of the new technological equipment, in order to act as advocates for its installation they might be tempted to make subjective estimates of the enhancement in efficiency that will result. In order to be numerical let us say that they have claimed (roughly) a fractional enhancement of efficiency of 10%; i.e. $\mathrm{d}h/h = 0.1$. Our first condition obtained from the "decision making" inequality shown above would be that the term inside the square brackets would *never* be negative. Therefore, using our hypothetical figure, annual costs of implementation should be less than (GNP/20).(0.1) which is 1/200th of the GNP.

Suppose, now, that the commercial vendors of the equipment know how to set the price of their equipment acceptably and have set it at 0.4% of the GNP. Then the value of the right-hand side of the inequality would be $(1/0.1)G(0.005-0.004) = 1/100$th of the GNP. Thus the discounted costs of research, development, and installation should not exceed 1/100th of the GNP.

It was stated earlier that examples involving quantification of social benefit which are believable and instructive are difficult to find. If the last example is not entirely believable I hope it has been instructive. Supposing that a similar model is again attempted I feel sure that some environmental disamenity will be subtracted from the social benefit and its magnitude will be related to the size of the GNP. However, such gambits have become the temporary province of Meadows (1972) in his work on the "The Limits to Growth" and my present task of recording information on tech-

160

nology assessment does not allow time for a digression on this interesting topic.

In all technology assessment, quantified analysis of the sensitivity of social benefit—cost should be our aim. Optimistically, I hope this is one field of endeavour for the application of science, which the younger generation will not throw out "with the bathwater" when rejecting too strong a dedication to the natural sciences. Success in this kind of analysis might only come to those who broaden their social awareness without sacrificing the need to become more analytically flexible than my own generation.

Comments on Mathematical Modelling in Benefit—Cost Analysis

The previous content of this chapter, particularly the examples which demonstrate benefit—cost models of off-shore mining and educational technology, illustrates the way in which analysis of anything other than a simple operation results in a mathematical model.

Semantically the word "model", as opposed to "theory", has much to recommend it to the analyst because it gives to the amathematical recipient a feeling, albeit a false one, of security. "Theory" to the mind of the practical decision maker can represent unnecessary elaboration of dubious outcome whilst modelling can represent something practical and useful. Naturally, not all models are practical or useful and a great deal of attention must be given to the possible usefulness and accuracy of any model that is to be used in technology assessment.

Spitz (1972) offers a prescriptive method for rating models which in use provides some guidance to the acceptability of modelling work. A summary of Professor Spitz's tests is as follows[1].

(1) Is the model devoid of symbolic formulae which may be more conveniently verbalized or omitted without loss of rigour?

(2) Is the model devoid of any serious shortcomings or limitations on the conceptualization of the subject matter?

---

[1] To exonerate myself partially from the consequences of applying these tests to the models in this book I feel obliged to point out that some of the first-order models described are selected more for instructive purposes than accuracy. However, as a heuristic exercise the reader should reflect on the quality of the models herein against a background of Spitz's tests.

(*3*) Is the model devoid of improper assumptions regarding relationships of scale or omission of variables?

(*4*) Can the required data be obtained and will processing the data reveal anything new or important?

(*5*) Is the model accompanied by a description of convincing test cases of the situation in which it should work?

When the answer to any of the above questions is "yes", Spitz recommends the award of a star, if the answer is "don't know" he suggests the award of half a star. The sum of the stars awarded is meant to give a nonmetric test of the model's quality in the way the military denotes the quality of generals! Such a scheme is obviously open to abuse but the questions do represent useful "rules-of-thumb".

In complicated analyses, simplification of a complex situation is the primary justification for mathematical modelling and it follows that careful simplification of a model which is intended to represent a real situation is highly desirable. The wood must be visible through the trees. Nevertheless, a model can be simplified improperly. Crucial parameters can be omitted and without these the model can hardly be expected to reveal its sensitivity to the omission. Alternatively, all crucial parameters can be included but without the intrinsic causal relationships between them.

When there is no a priori knowledge of the causal relationships, they are sometimes replaced with subjectively chosen relationships which have, or have not, tenuous justification in known experience.

All in all, the path recommended by Bellman (1957), when referring to a specific type of modelling between the "pitfalls of over-simplification and the morass of over-complication", is not negotiable without stepping through some very difficult terrain! Mathematically competent recipients of models can develop their discernment in the perusal of models by assuming, until conviction (or otherwise) of the models acceptability dawns, that the modeller's plausability exceeds his capability. Whatever time is consumed in checking the mathematical assumptions made, it is usually much less than that expended by the modeller in creating the model.

Amathematical recipients face a difficult task of assessing the model's validity only when they are not the arbiters-of-decision and have no real authority to counterbalance one group of modellers with another. The amathematical arbiter often solves his problem by staging a debate between protagonists of different modelling approaches. Indeed, this last approach has recently become so popular in some quarters that arbiters call for preliminary models at the onset of an assessment project and then use the outcome from initial debates between modellers as an indication of how the project should be managed when it is not amenable to verbal and "literary" approaches.

The present pre-eminence of amathematical arbiters-of-decision is not permanent and cannot, if the present belief in quantification is maintained, continue. Commitment to more "scientific" approaches to planning and assessment necessarily predicates an increasing competence amongst arbiters-of-decision in the use of mathematical technique and an apposite one-sentence scenario to this effect has been made by a distinguished American (who shall remain anonymous): "In twenty years time the mathematical modellers of psychological processes will be the chief advisors to the President". Whatever skills in modelling are acquired in the future, there will always be a basic dilemma in the use of some kinds of "predictive" model which attempts to simulate the dynamics of change. Normally, as has been indicated, the construction of a worthwhile model involves iterative testing of its accuracy but a model of a changing system cannot be tested until *after* the charge is partly complete. It is, therefore, valid to use modelling in long-range analysis, when ex post and/or incremental checks in the near-future are not sufficient validation? There is no definitive answer to this question because all long-range "predictive" models are subjective and should be regarded as a class of models whose *main* purpose is to provide a more effective means of discussing the future. No denigration is meant in relegating such a class of models from semi-accurate predictors to the category of communication channels, for they do at least deal more effectively (if you understand the language) with large numbers of parameters than non-mathematical, verbal, forms of communication. One pleasing aspect of the present trend is that amathematical generalists regard modelling more and more as a necessary adjunct to discussion. (Modelling in technological forecasting is discussed at greater length in Chapter 7.)

APPENDIX 1

Present Values of Cash Flows

This appendix complements Chapter 5, on benefit—cost analysis and contains two parts, concerned exclusively with continuous methods of discounting.

(1) A section giving in algebraic form the present values of commonly occurring general analytic shapes of cash flow curve, and

(2) a section which explains how continuous methods of discounting can be easily applied to non-analytic curves, defined solely by graphs and/or numerically.

Analytic Cash Flow Curves

By using the formula for the present value, obtained with continuous methods of discounting, it is possible to derive meaningful algebraic formulas for present values. The procedure is simply to put the formula for the cash flow $d£/dt$ into the integral $\int_0^\infty (d£/dt)e^{-rt}\,dt$ and integrate.

As was said before, there is often no need to puzzle too long over the solution to the integral because there are copious tables of Laplace Transforms (which is what the integrals are) in the mathematical, scientific and engineering literature (see, for example, Abramowitz and Stegun (1965) p. 1021, or Jaeger (1951) pp. 3, 24 or 126).

The simplest example is a cash flow, constant at a level of $£A$ per annum, starting at time zero and extending an infinite time into the future, *viz.*

The present value of this cash flow is obviously $£\int_0^\infty A\,e^{-rt}\,dt$ which integrates to $£(A/r)$.

The next simplest, excluding the case when $(d£/dt)$ is negative and constant when the present value is $-£(A/r)$, is a constant cash flow which starts at some time T (not equal to zero) and extends for an infinite time, *viz.*

164

The present value is $\pounds\int_T^\infty A e^{-rt}\, dt$ which integrates to $\pounds (A/r) e^{-rT}$. Note that the effect of delaying the cash flow by T years is to reduce the cash flow by a multiplying factor e^{-rT} which is always less than unity[1].

In general, whatever the shape of the cash flow, if $\overline{\pounds(r)}$ is the present value when the cash flow starts at zero, then $e^{-rT}\overline{\pounds(r)}$ will be the present value if the same shape of cash flow starts at T. It can be understood another way by first considering that if the cash flow is discounted to its start point, then its value discounted to time T would be $\overline{\pounds(r)}$ and this value discounted to zero time would be $\overline{\pounds(r)}$ multiplied by e^{-rT}, i.e. $e^{-rT}\overline{\pounds(r)}$.

Using the last result, it is not too difficult to discern that a cash flow, constant at $\pounds A$ p.a., between T_1 and T_2 has a present value of

$$\int_{T_2}^{T_1} A\, e^{-rt}\, dt = A(e^{-rT_1} - e^{-rT_2})/r$$

because the result may be obtained by superposition, i.e. simply subtracting the present values of two cash flow curves that yield the cash flow curve required.

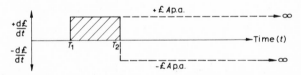

In assessing the likelihood of growth, or decay, a common first assumption is that the change in a cash flow will be linear. Therefore, it is useful in present value calculations to have algebraic formulas for "ramp" like changes. Consider a linear growth of cash flow starting at time zero and growing in time T to a constant level of $\pounds A$ p.a.

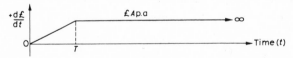

---

[1] This is a very simple example of the Heaviside Shift Theorem (see Jaeger, 1951, p. 82). For those not allergic to mathematical notation, if $\overline{\pounds(r)}$ is the present value of $d\pounds/dt = f(t)$ and $T > 0$, then $e^{-rT}\overline{\pounds(r)}$ is the present value of $U(t-T)f(t-T)$ where $U(t) = 0$, $t < 0$ and $U(t) = 1$, $t > 0$.

165

The present value of this cash-flow is

$$\pounds \int_0^\infty \frac{d\pounds}{dt}\, e^{-rt}\, dt = \pounds \int_0^T \frac{tA}{T}\, e^{-rt}\, dt + \left(\frac{A\, e^{-rT}}{r}\right)$$

Obviously the first integral is not too difficult to solve but useful heuristic exercise may be got from applying the Shift Theorem to its solution.

Tables of Laplace transformations show that if the "ramp" had continued to infinity, its present value (the integral's value if the upper limit equals ∞) will be (A/r^2). And a "ramp" which drops to zero at time T may be got from an infinitely long ramp minus a similar ramp starting at T minus a constant cash flow also starting at time T as follows.

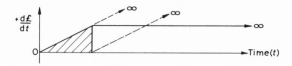

Thus using the Shift Theorem, the present value of the shaded area will be

$$\frac{A}{T}\frac{1}{r^2} - \frac{A}{T}\frac{1}{r^2}\, e^{-rT} - \frac{A}{r}\, e^{-rT}$$

Using this formula and adding the contribution for the plateau in the cash flow we started with, it is clear that the present value of a ramp followed by an infinite plateau will be

$$\frac{A}{T}\frac{1}{r^2}(1 - e^{-rT})$$

Masochistic readers are welcome to work out the complicated series which would result if the present value is computed by discontinuous methods of discounting and compare it with the conciseness of the last result!

A cash flow consisting of a ramp of width T starting at time T_1 followed by a plateau starting at time $(T_1 + T)$ and continuing to infinity would have a present value using the Shift Theorem, equal to the last result multiplied by e^{-rT_1}, i.e.

166

$$\frac{A}{T}\frac{1}{r^2}e^{-rT_1}(1-e^{-rT})$$

Obviously, if the plateau did not continue for infinite time but was truncated with a vertical cut at time T_2 (say) we would need to subtract $(A/r)e^{-rT_2}$ from the above present value.

In order to indicate the method of dealing with a negatively sloping ramp, the present value of the following shaded cash flow will be computed.

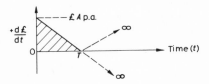

The shaded area may be obtained by the superposition of a negative ramp $[A-(At/T)]$ starting at $t=0$ plus a positive ramp starting at $t=T$ and continuing to infinity. The first part plus the second part gives a present value of

$$\frac{A}{r}-\frac{A}{r^2}\frac{1}{T}+\frac{A}{r^2}\frac{1}{T}e^{-rT}$$

i.e.

$$\frac{A}{r}-\frac{A}{r^2}\frac{1}{T}(1-e^{-rT})$$

This, incidentally, is also the present value of a constant plateau starting at zero minus a ramp starting at zero and followed by an infinite plateau, thus showing a different superposition for achieving the same result.

In the table of cash flows and associated algebraic present values which follows this section of the appendix, the reader will find many shapes of cash flow comprised of ramps and plateaux. Enough has been said in the preceding examples to allow the independent verification of the results given by the author.

Finally, before proceeding to read the tabular presentation of present values, it is useful to consider the case of a discrete amount of cash, £S (say), acquired at time T. Such an amount on

TABLE 1

Tabular presentation of some commonly occurring cash-flows

Shape of cash flow curve	Present value of cash-flow (continuously discounted)
Single payment 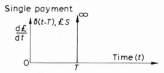	$£S.e^{-rT}$ r is the discount-rate
Constant rate 	$£A/r$
Truncated constant rate 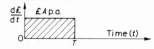	$£(A/r).(1-e^{-rT})$
Delayed constant rate 	$£(A/r).e^{-rT}$
Delayed and truncated constant rate 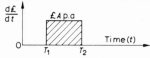	$£(A/r).(e^{-rT_1}-e^{-rT_2})$
Constant rate approached by ramp 	$£\dfrac{A}{Tr^2}.(1-e^{-rT})$
Delayed ramp + plateau 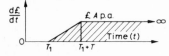	$£\dfrac{A}{Tr^2}.(1-e^{-rT}).e^{-rT_1}$
Truncated ramp 	$£\left[\dfrac{A}{Tr^2}.(1-e^{-rT})-\dfrac{A}{r}.e^{-rT}\right]$

168

Negative ramp

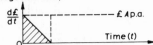

$$\mathcal{L}\left[\frac{A}{r} - \frac{A}{Tr^2} \cdot (1 - e^{-rT})\right]$$

Constant cost followed by constant benefit

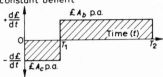

$$\mathcal{L}\left[\frac{A_b}{r} \cdot (e^{-rT_1} - e^{-rT_2}) - \frac{A_c}{r} \cdot (1 - e^{-rT_1})\right]$$

and the benefit—cost ratio is

$$\frac{\overline{B(r)}}{\overline{C(r)}} = \frac{A_b \cdot (e^{-rT_1} - e^{-rT_2})}{A_c \cdot (1 - e^{-rT_1})}$$

Unbounded exponential growth

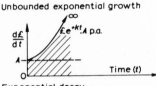

$$\frac{\pounds A}{(r - k)} \quad but \text{ only when } k \text{ is less than } r.$$

Exponential decay

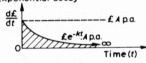

$$\frac{\pounds A}{(r + k)}$$

Bounded exponential

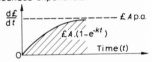

$$\mathcal{L}\left[\frac{A}{r} - \frac{A}{(r + k)}\right]$$

Trapezoid

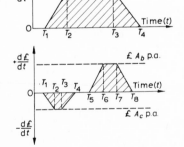

$$\frac{\pounds A}{r^2} \cdot \left[\frac{(e^{-rT_1} - e^{-rT_2})}{T_2 - T_1} - \frac{(e^{-rT_3} - e^{-rT_4})}{T_4 - T_3}\right]$$

$$\left\{\frac{\pounds A_b}{r^2}\left[\frac{(e^{-rT_5} - e^{-rT_6})}{T_6 - T_5} - \frac{(e^{-rT_7} - e^{-rT_8})}{T_8 - T_7}\right] - \right.$$

$$\left.\frac{\pounds A_c}{r^2}\left[\frac{(e^{-rT_1} - e^{-rT_2})}{T_2 - T_1} - \frac{(e^{-rT_3} - e^{-rT_4})}{T_4 - T_3}\right]\right\}$$

The benefit—cost quotient is obviously the first part of the present value divided by the second.

Half sine wave

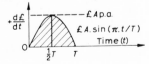

$$\mathcal{L}\,\frac{A\pi \cdot T}{r^2 T^2 + \pi^2} \cdot (1 + e^{-rT})$$

a cash flow curve would look like an infinite spike, i.e. an infinite rate of flow of cash occurring in an infinitesimally small time (recognizable to the mathematically minded as a Dirac spike of area £S). The non-specialist need, however, have no difficulty with discounting such an unusually shaped cash flow because its present value is simply $£Se^{-rT}$ (see the equation used in the explanation of continuous method which appeared just before the first integral on p. 129).

Numerical Procedure for Calculating Net Present Value of Non-analytic Cash Flow Curves

The formulas given in the previous section are only of use when the actual cash flow approximates to one of the analytic forms which are listed. Fortunately, most cash flows commonly occurring in analysis do approximate to the listed, or a combination of the listed, shapes. However, occasionally non-analytic shapes need to be discounted and recourse must then be made to numerical analysis. (When this happens it is a good idea to challenge the derivation of the shape of the cash flow curve because unduly "accurate" and "sophisticated" shapes are not usually commensurate with the uncertainty of the situation.) In this section of the Appendix a useful numerical method for approximating to the present value is demonstrated.

Suppose the project assessment procedure yields a cash flow curve, denoted by the solid line in the following diagram.

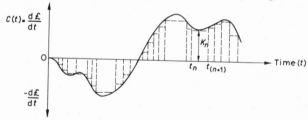

We first approximate to the curve by choosing a number, N, of rectangles of varying heights and varying widths (in time). The nth rectangle will have height K_n (+ ve or − ve), be bounded by ordinates t_n and $t_{(n+1)}$, and can be described by [1]

---

[1] The symbols of type $U(t-T)$ are simply mathematical shorthand to indicate the sudden steps, up and down, at the beginning and end of the rectangle.

$$c_n(t) = K_n[U(t-t_n)-U(t-t_{(n+1)})]$$

where $U(t-T) = 0$, $t < T$; $U(t-T) = 1$, $t > T$.

As in the previous section, we see that the net present value of the nth rectangular distribution of cash flow is

$$P_n(r) = \frac{K_n}{r}\left[\exp(-t_n r) - \exp(-t_{(n+1)}r)\right]$$

Since the net present value of the total distribution is

$$P = \sum_{n=1}^{n=N} P_n(r)$$

we deduce that

$$P = \sum_{n=1}^{N} \frac{K_n}{r}\left[\exp(-t_n r) - (-t_{(n+1)}r)\right]$$

Furthermore, if we sum separately over all negative values of K_n and then all positive values of K_n, we find

$$\text{Discounted benefit} = B = \sum_{n=q}^{N} \frac{K_n}{r}\left[\exp(-t_n r) - \exp(-t_{(n+1)}r)\right]$$

$$\text{Discounted cost} \quad = C = \sum_{n=1}^{q-1} \frac{K_n}{r}\left[\exp(-t_n r) - \exp(-t_{(n+1)}r)\right]$$

where the rectangles up to and including $n = (q-1)$ are negative.

To find the inherent effective rate of return it is first necessary to find the inherent nominal rate of return equating B to C. This will not, in general, yield an analytic solution for R and numerical methods will be necessary.

An Example of the Use of the Numerical Method

For the purpose of this example we assume that the arbiters of decision authorized a technological project which has had the cash

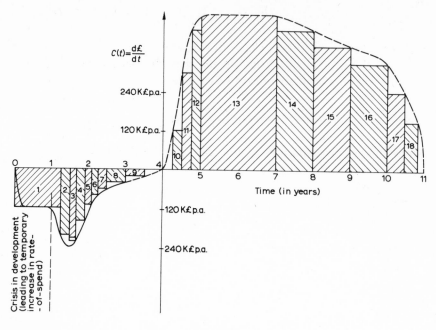

Fig. 10.

flow curve shown by the full line on the graph in Fig. 10. We also assume that the latest extrapolations of the cash flow to date indicate that the likely outcome will be given by the dotted line. We then ask ourselves what the expected net present value and the benefit—cost ratio referred to time zero will be for a nominal discount rate of 10%.

The first step in the calculation is obviously to approximate to the cash flow characteristic with rectangular blocks. This has been done on the graph. Each shaded region represents a rectangle. There are nine rectangles of negative K_n, the first nine entries in Table 2, and nine rectangles of positive K_n which are the last nine entries[1].

The nominal rate of discount is $r = 0.1$ and the values of K_n, t_n,

---

[1] A quick perusal of the rectangular shapes will show that they are of very different widths, narrow when accuracy is desired and wide when the fluctuations of the cash flow curve are slow. Using discontinuous methods of discounting, the areas would be of uniform width, appropriate to the accounting period or the analyst would have to remember different rules of discounting for sums accrued over different periods. Furthermore, he would have to calculate areas under the curve.

172

TABLE 2

Itemized discounted cash flow
(K_n in £p.a., t in years, r = 0.1 and $S_n = K_n[\exp(-t_n r) - \exp(-t_{(n+1)} r)]$)

Rectangle no.	K_n	t_n	$t_{(n+1)}$	$\exp(-t_n r)$	$\exp(-t_{(n+1)} r)$	S_n
1	-1.2×10^5	0	1.25	1	0.88	-1.44×10^5
2	-2.0×10^5	1.25	1.5	0.88	0.86	-0.4×10^5
3	-2.1×10^5	1.5	1.58	0.86	0.85	-0.21×10^5
4	-1.6×10^5	1.58	1.92	0.85	0.825	-0.4×10^5
5	-1.1×10^5	1.92	2.08	0.825	0.812	-0.14×10^5
6	-0.8×10^5	2.08	2.25	0.812	0.800	-0.096×10^5
7	-0.6×10^5	2.25	2.5	0.800	0.779	-0.12×10^5
8	-0.4×10^5	2.5	3.0	0.799	0.741	-0.23×10^5
9	-0.2×10^5	3.0	3.5	0.741	0.705	-0.07×10^5
						-3.11×10^5
10	$+1.2 \times 10^5$	4.25	4.5	0.654	0.638	$+0.19 \times 10^5$
11	$+3 \times 10^5$	4.5	4.75	0.638	0.622	$+0.48 \times 10^5$
12	$+4.3 \times 10^5$	4.75	5.0	0.622	0.607	$+0.65 \times 10^5$
13	$+4.8 \times 10^5$	5.0	7.0	0.607	0.497	$+5.4 \times 10^5$
14	$+4.3 \times 10^5$	7.0	8.0	0.497	0.449	$+2.06 \times 10^5$
15	$+3.8 \times 10^5$	8.0	9.0	0.449	0.407	$+1.6 \times 10^5$
16	$+3.3 \times 10^5$	9.0	10.0	0.407	0.368	$+1.32 \times 10^5$
17	$+2.4 \times 10^5$	10	10.5	0.368	0.350	$+0.43 \times 10^5$
18	$+1.5 \times 10^5$	10.5	1.83	0.350	0.339	$+0.17 \times 10^5$
						$+12.3 \times 10^5$

173

and $t_{(n+1)}$ for each rectangle are taken from the graph and shown in Table 2 together with the relevant terms in the series[1] listed above. Each term in the series is denoted by

$$S_n = (K_n/r)[\exp(-t_n r) - \exp(-t_{(n+1)} r)]$$

Summation of the negative values of S_n yields the discounted cost, i.e.

$$\overline{C(r)} = -\pounds 3.11 \times 100,000$$

Similarly, summation of the positive values of S_n gives the discounted benefit, i.e.

$$\overline{B(r)} = +\pounds 1.23 \times 1,000,000$$

Then

$$\frac{\overline{B(r)}}{\overline{C(r)}} = 3.96$$

and

Present value = £920,000

APPENDIX 2

A Detailed Example of the Use of Sampling Function in Benefit—Cost Analysis

This appendix complements that part of Chapter 5 dealing with continuous probability distributions. When continuous distributions are used to indicate the likely distribution of benefit and cost, analysts are faced with combining probability density functions in such a way that an overall sampling distribution is obtained. In this example it is assumed that the criterion used in project assessment is the quotient of benefit—cost and the PDFs of benefit and cost are assumed to be rectilinear distributions.

---

[1] Terms of kind e^{-tr} are easily obtained from tables of exponential functions which list e^{-x} (see, for example, Abramowitz and Stegun, 1965, p. 116).

174

Three possible overall sampling distributions are deduced, the shape of each being dependent upon the cut-off values of the benefit and cost distributions. The mean and the variance of each distribution are algebraically derived.

Statistical Analysis

Following the nomenclature of Kendall and Stuart (1963), the sampling distribution of the ratio of independent variates x_1 and x_2 is written as

$$dF(x_1, x_2) = dF_1(x_1)dF_2(x_2) = f_1(x_1)f_2(x_2)dx_1dx_2 \qquad (1)$$

and the marginal distribution of $u = (x_1/x_2)$ is

$$dH(u) = \int_{-\infty}^{+\infty} f_1(ux_2)f_2(x_2)x_2\,dx_2\,du \qquad (2)$$

The last integral is reasonably straightforward when the distributions f_1 and f_2 do not possess discontinuities. A cogent example where f_1 and f_2 are normal distributions is given in Kendall and Stewart (1963) Vol. 1. But as was explained in Chapter 5, benefit—cost analysis is often carried out under exigencies which lead some analysts to assume rectilinear distributions for f_1 and f_2.

Consider integral (2) for the case when f_1 and f_2 are rectilinear with lower and upper cuts-off of (a, b) and (c, d) respectively. Then the heights of f_1 and f_2 distributions, obtained from the normalizing condition that the area under the PDF shall be unity, are as follows.

$$k_1 = (b-a)^{-1} \quad \text{and} \quad k_2 = (d-c)^{-1} \qquad (3)$$

and integral (2) yields

$$dH(u) = \tfrac{1}{2}k_1k_2\left[x_2^2\right]_{\text{maximum of }(c,\,a/u)}^{\text{minimum of }(d,\,b/u)} \qquad (4)$$

where

$$a/d < u < b/c \qquad (5)$$

175

Thus the upper limit of integral (2) is b/u for $b/d \leqslant u \leqslant b/c$ and d for $a/d \leqslant u \leqslant b/d$, and the lower limit is c for $a/c \leqslant u \leqslant b/c$ and a/u for $a/d \leqslant u \leqslant a/c$. Since the ratio of the upper cut-offs may be greater than, less than, or equal to the ratio of the lower cut-offs, there are three general cases for the marginal or overall sampling distribution.

Case I. $b/d > a/c$

dH(u)	Range
$\frac{1}{2}k_1 k_2 (d^2 - a^2/u^2)du$	$a/d < u < a/c$
$\frac{1}{2}k_1 k_2 (d^2 - c^2)du$	$a/c < u < b/d$
$\frac{1}{2}k_1 k_2 (b^2/u^2 - c^2)du$	$b/d < u < b/c$

Case II. $a/c > b/d$

dH(u)	Range
$\frac{1}{2}k_1 k_2 (d^2 - a^2/u^2)du$	$a/d < u < b/d$
$\frac{1}{2}k_1 k_2 [(b^2 - a^2)/u^2]du$	$b/d < u < a/c$
$\frac{1}{2}k_1 k_2 (b^2/u^2 - c^2)du$	$a/c < u < b/c$

Case III. $a/c = b/d$

dH(u)	Range
$\frac{1}{2}k_1 k_2 (d^2 - a^2/u^2)du$	$a/d < u < a/c = b/d$
$\frac{1}{2}k_1 k_2 (b^2/u^2 - c^2)du$	$a/c = b/d < u < b/c$

The cases I, II, and III are all sketched in Fig. 5, p. 144.

All the above sampling distributions, of course, integrate to unity and the first moment about $u = 0$ for all cases is

First moment about zero = mean = $[\frac{1}{2}(a + b)/(d - c)]\ln(d/c)$ (6)

176

Therefore,

$$\frac{\text{Mean of the benefit—cost ratio}}{\text{Ratio of means of benefit and cost}} = \frac{1}{2} \frac{(d/c) + 1}{(d/c) - 1} \ln(d/c) \qquad (7)$$

The second moment about the point $u = 0$ in all cases is

$$\text{Second moment about zero} = \frac{a^2 + ab + b^2}{3cd} \qquad (8)$$

The variance is the second moment about zero minus the square of the mean, i.e.

$$\text{Variance of benefit—cost ratio} = \frac{a^2 + ab + b^2}{3cd} - \frac{1}{4} \frac{(a+b)^2}{(d-c)^2} \ln^2(d/c)$$

$$(9)$$

References

Abramowitz, M. and Stegun, Irene A. (1965). *Handbook of Mathematical Functions*, New York: Dover Publications, Inc.

Bellman, R. (1957). *Dynamic Programming*, Princeton University Press.

Dupuit, J. (1844). *On the Measurement of Utility of Public Works*, International Economic Papers, Vol. 2.

Jaeger, J.C. (1951). *An Introduction to the Laplace Transform*, London: Methuen.

Katz, M. (1970). *Hearings before the Subcommittee on Science, Research, and Development, of the Committee on Science and Astronautics, U.S. House of Representatives, 91 Congress*, Washington: U.S. Govt. Printing Office.

Kendall, M.G. and Stuart, A. (1963). *The Advanced Theory of Statistics*, London: Charles Griffin.

Mansfield, Edwin *et al.* (1972). *Research and Innovation in the Modern Corporation*, London: The MacMillan Press.

Medford, R.D. (1969). "Marine Mining in Britain", *Mining Magazine*, 121, Nos. 5 and 6.

Meadows, Dennis L. *et al.* (1972). *The Limits to Growth*, A Report of the Club of Rome's Project on the Predicament of Mankind, Washington: Potomac Associates.

N.C.E.T. (1969). *Computer-Based Learning Systems*, London: National Committee for Educational Technology.

Prest, A.R. and Turver, R. (1965). "Cost—Benefit Analysis: A Survey", *The Economic Journal*, December 1965.

Spitz, Edward A. (1972). "A Prescriptive Method for Rating Models", *Industrial Marketing Management*, 1, No. 3, 358.

CHAPTER 6

Cost—Effectiveness Analysis
and Technology Assessment

"What is a cynic?"

*"A man who knows the price of everything
and the value of nothing."*

Oscar Wilde

Introduction

There is an explicit assumption made in the previous chapter on benefit—cost analysis that, no matter how difficult it may be, the benefit can be quantified. Exigencies of real life, however, often force the analyst into a situation where, within the available time, the costs are tangible but the benefits are not. In such a situation, where the benefits are not really quantifiable but are not qualitatively in question, the analyst is constrained to use a proxy measure of benefit which is loosely described as the "effectiveness" of the programme. All the stratagems of benefit—cost analysis may be used when the goal is to be aimed at regardless of the unknown value of the quantified benefit, but the benefit is measured in the proxy units.

Uncomfortably crude measures of the proxies of benefit, taken from military systems analysis, are contained in the following cost—effective ratios: the expected number of enemy killed per

178

unit of weapon-system cost, the expected quantity (men X miles/time) a transportation system can achieve, at a defined level of vulnerability, per unit of cost, and the explosive equivalent of a nuclear weapon, in tons of TNT, per unit cost. More comforting examples are the increment in first-class protein that the richer countries can supply to their less prosperous neighbours per unit of altruistic aid, the non-disruptive throughput of a civil transportation system per unit of cost and the energy equivalent, in tons of coal or fossil fuel, a nuclear power station can produce per unit of real cost (including pollution).

All measures of proxy benefit, or effectiveness, are subject to criticism as, indeed, are all quantifications of "measurable" benefit. The art of analytic refinement lies in the selection of acceptable parametric measures of effectiveness. And the combination of *all* relevant factors into an overall measure of effectiveness involves much finesse in the acceptable juxtaposition of the "good" and "bad" outcomes of any programme analysed. For example, in the definition of the non-disruptive throughput of a motorway, it would be necessary to arrange the disruptive factors like air pollution, noise, and aesthetic offence so that high values of these quantities would adversely affect the measure of effectiveness. But the arrangement, or counterbalancing, of such factors depends critically upon the E.Q. (ethical coefficient) of the analyst or the arbiter-of-decision, see Gabor's suggestion reported in Chapter 7. Therefore, there will never be unanimous agreement about the acceptability of measures of effectiveness and the public must be very cautious about accepting those measures which appear to have the plausibility of propaganda slogans. In benefit—cost analysis it is relatively easy to notice that some crucial disbenefit has not been included in the calculation. But in cost—effectiveness analysis it is not so easy to notice that the parametric representation of a disbenefit does not give it sufficient weight. A classic case of this may be the often quoted American example where the acquisition of air-conditioning by one sector of the population creates more discomfort, by electrically generating more total heat, for that part of the population without air-conditioning. Thus a measure of environmental comfort in sweltering climes may not be a straightforward measure of air-conditioners per capita of population.

Unless the goal to be achieved is a highly personal one (see the attempt at a humorous cost—effectiveness calculation at the end of this chapter), effectiveness must be a measure of collective utility to an organization or a society and this involves aggregates of individual utilities. As the general subject under discussion is technology assessment it is appropriate that the example of cost—effectiveness used in this chapter should include an economist's concept of sub-group and central utilities and also deal with a recognized technological innovation, the digital computer, which is a force of change.

The analysis presented below stemmed from the work of the Programmes Analysis Unit. It is previously unpublished work of F. Walford and myself. Due to the reorganization of the old Ministry of Technology when it was transformed into the Department of Trade and Industry, the method of analysis described was not applied in the public service. The example should, therefore, be regarded as an aid to exposition of some of the principles underlying calculation of cost—effectiveness. Any views expressed about the method or its application are purely my own and should not be used to infer departmental procedures.

An Example of the Cost—Effectiveness of Computers

An increasing proportion of the resources of technical ministeries throughout the world is being absorbed by the provision of extensive computer facilities. This example is based on work carried out to see whether any further light can be shed on the problem of identifying and, wherever possible, measuring the benefits obtainable from the use of computers in large technological and scientific research programmes.

In essence the approach is an extension of the work of Smidt (1968). It is assumed that each leader of a sub-group has optimized the usefulness of his group by optimally arranging his resource allocations between computing and non-computing activities within the budget constraint imposed by the central organization and the limitations of existing computer technology, performance of himself and his staff, and the inherent nature of the work in hand. A further assumption is that the central organization has made an optimal allocation of budgets to the sub-groups in order to maximize the overall utility derived by the central organization. In economist's nomenclature this means that a Pareto optimization of central utility has been made subject to certain constraints. Parametric representation of effectiveness arises, in this example, in the following ways.

180

(*1*) How is the efficacy of the sub-group represented as a function of the parameters which go towards success or failure?

(*2*) How is the central utility represented as a function of the sub-group utilities?

The objectives of carrying out a cost—effectiveness analysis are determined by customer needs. Specifically this analysis was commissioned to help decision making on the provision of computer facilities within a very large organization. Intentions were that the analysis could be used to determine what advantage, or disadvantage, would accrue from the installation of more technologically advanced computers, what the propensity to spend on computers is for a specific type of R & D, whether a central computer, with on-line facility is preferable to decentralized computers, and, finally, if by changing accounting procedures (used to charge for use, internal to the organization, of the organization's computers) the existing organization, for the provision of computer facilities, could be beneficially perturbed.

Utility Functions and Smidt's Model

Complete regurgitation of Smidt's economic models for the effective control of computer usage in an organization whose objectives cannot be readily expressed in a monetary terms is a lengthy procedure. For present purposes, an abbreviated verbal description of one of his models should suffice.

Assume that all leaders of sub-groups receive funds from the central organization and that one leader cannot transfer funds to, or from, another. Two commodities, in the following categories, are available to the leaders.

Commodity 1, representing computer services, is produced by the organization and decisions about the quantity and quality available, and the price at which it will be sold, are made by the organization;

Commodity 2, representing labour, material, and equipment (other than computers and the paraphernalia of computing) is purchased from outside the organization at a price which cannot be effectively changed by the behaviour of the central organization or by the collective or individual behaviour of its sub-group leaders.

For the leaders it is useful to think of the price of computing as given and for the organization to think of the price as a policy variable with computer capacity as something that is determined by the amount demanded at that price.

181

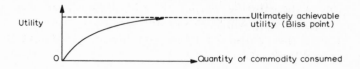

Fig. 1. A utility curve.

Leaders will try to maximize the effectiveness, hereafter called utility, of their sub-groups. The customary requirements of any utility function are that it should satisfy the laws of diminishing returns[1] which, in geometrical terms, are that the shape of the utility curve should be concave as viewed from the axis representing quantity of commodity consumed (see Fig. 1).

The constraint upon the leader is that the total cost of the quantities of the two commodities he uses should not exceed his available income, determined by the central budget. In satisfying this constraint, and the condition that the utility curve must be concave to both kinds of commodity consumed, he arrives at the condition that the marginal enhancement of utility per pound spent on computing should equal the marginal utility per pound spent on non-computing. And that each of these quantities equals the marginal increment of the sub-group's utility per pound of budget allocated from central funds.

The objective of the central organization is to maximize the overall utility which is some function of the sub-group's utilities. Unless the central organization has some knowledge of causality which predetermines otherwise, it usually computes, consciously or subconsciously, the total utility by simply adding weighted sub-group utilities. (If more complicated functions of sub-group utilities are used they should, of course, not allow the possibility of goal conflict between sub-groups.) In maximizing the central organization's utility, the budget constraint is that the amount allocated to the users is the sum of the fixed resources plus the income from the computing time sold to the sub-groups and an optimum allocation of budgets to the sub-groups is one in which the ratio of the marginal utility derived by the organization per pound spent on any user to the corresponding net budget drain is the same for all other users.

---

[1] Mathematical readers will recognize this requirement as one which constrains the first and second derivates of utility, with respect to amount of commodity consumed, to be respectively positive and negative.

182

If analysis is further used to solve the problem of optimal price setting, it is possible to prove that the computer services should be priced at their marginal cost to the organization, in which case the net budget drain is zero.

Unfortunately, Smidt does not assume any particular form of utility function but despite this he draws some interesting conclusions about the cases where the price of computer services is greater, and less, than their marginal cost. He also reveals the consequences of making larger, and smaller, budget allocations to leaders with a high, and low, propensity to spend money on computing.

An Extension of Smidt's Model

Smidt's model only gives an indication of how one would set about a cost—effectiveness analysis of the efficacy of computing in, and the impact of new computer technology on, R & D activities which in themselves are likely to yield relatively unpredictable results. One major hurdle to be got over before the application of his model is that each specific R & D activity, which utilizes, to some degree, the services of the computer has a different utility function. Some R & D generates a very large propensity to use the computer, whilst other kinds of R & D inherently need computing support very little. Research on boundary layer problems in aerodynamics, say, would be expected to make much use of the computer but experimental research into new kinds of physical effects may not.

So that a start could be made to the analysis it was necessary to postulate a sub-group utility of the following form

$$\text{Utility} = u = [1 - \exp(-k_1 S_1)] . [1 - \exp(-k_2 S_2)] \qquad (1)$$

where S_1 and S_2 are the spends on computing and non-computing and k_1 and k_2 are constants which determine the intrinsic need for computing and non-computing commodities. This is merely a multiplicative form[1], for two commodities, of a very standard and

---

[1] This form satisfies the requirement that the utility function should be concave with respect to either commodity. (The law of diminishing returns with respect to each commodity is obeyed.)

An additive form of utility function such as

$$\text{Utility} = A[1 - \exp(-k_1 S_1)] + B[1 - \exp(k_2 S_2)]$$

183

frequently used utility function of one commodity. It has the advantage that the maximum utility obtainable is unity and that sub-group activities with infinitesimal inherent need for computing (or non-computing commodities) have a value of k_1 (or k_2) which is infinite ($\exp(-kS)$ tends to zero as kS tends to ∞). A large inherent need for either commodity is, in contrast, denoted by a small value of the constant k, which means that there must be a relatively large spend on the commodity to achieve reasonable values of utility.

In an ideal analyst's world the leaders would know what the value of k_1 and k_2 were for their sub-group's research, and the cost—effectiveness calculations would then become trivial. In reality, most leaders "fly by the seat of their pants" and make allocative decisions within their sub-groups by methods which are a mixture of calculation and intuition. (Management is always the assessor of performance in its own field.) This does not mean, however, that techniques should not be adduced for the evaluation of values of the "constants" k which approximate to the leader's and his sub-group's behaviour[1]. Central management might also

(footnote continued from previous page)

is not appropriate for it would imply that the computing and non-computing activities were not as intertwined as experience indicates they are in the research situation. In this utility function, zero-spend on computing would still give a utility equal to some fraction of B.

In contrast, to get a non-zero utility with the multiplicative, form of utility function, one must spend on *both* commodities, even though the spend on either need not be large. (The model allows for non-digital computing by inserting a different value of k_1.) Furthermore, the additive form of utility function has one extra constant A (or $B = (1-A)$ since the normalizing condition is that the maximum utility equals unity).

The above reasoning does not apply to the additive form used below for the central utility. The reason is that the contributions from widely differing research programmes are not all essential to the organization's goal. One might be very effective overall with a successful biological project even though one fails miserably with a project on astrophysics: the two are not, essentially, mutually dependent.

[1] The "constants" k are only truly constant for invariant conditions. If changes take place in the leadership or staff of the sub-group, the technology or science of the subject under study, and experimental or computing technique, the value of the k's will change. For example, if *necessary* calculations are done laboriously by hand, they will be expensive and k_1 will be small but if the sub-group acquires a member who can facilitate the use of the computer to do the same calculations, the value of k_1 will increase. Conversely, the same newcomer might point out that the use of the computer was unnecessary and achieve an enhancement of k_1 in a different way.

"fly by the seat of its pants" but it has less justification in doing so because of the poorer feedback through the extended communications and the more disastrous consequences of a bad decision. The use of utility functions and the concomitant "theory" does not usurp the responsibilities of management at any level nor does it presume to evaluate their technical performance. It simply helps management systematize the available, but otherwise non-cohesive, information through the use of a fairly tenable hypothesis.

When the leader optimizes his sub-group's utility by equating marginal returns to a unit of spend on either commodity he is, in terms of the utility function above, saying

$$k_1 x(1-y) = k_2 y(1-x) \tag{2}$$

where, for algebraic convenience, x has been substituted for $\exp(-k_1 S_1)$ and y for $\exp(-k_2 S_2)$.

Thus the optimum utility of the sub-group obtained by substituting eqn. (2) in eqn. (1) is

$$u^* = \frac{(1-x)^2}{(m-1)x + 1} \tag{3}$$

where m is (k_1/k_2).

Equation (2) also determines the ratio of spend on computing (S_1) to the total sub-group budget (I) [1].

$$\frac{S_1}{I} = \frac{\ln(1/x^{\frac{1}{m}})}{\ln[(m-1)x + 1] - \ln(mx^{(1+\frac{1}{m})})} \tag{4}$$

From this last relationship, the propensity to spend on computing is easily derived (i.e. how much of a budget increment dI is this sub-group likely to spend on computing, dS_1).

$$\frac{dS_1}{dI} = \frac{(m-1)x + 1}{(m-1)x + (m+1)} \tag{5}$$

---

[1] Amathematical readers should not try too punitively to derive any of the following formulae. It is enough to note that they can be derived and see which quantities are involved in each formula.

Similarly, the marginal increment of sub-group utility per unit of budget increment is

$$\frac{du^*}{dI} = \frac{k_2 mx(1-x)}{(m-1)x + 1} \tag{6}$$

and the ratio of fractional change of utility to fractional change of budget (i.e. the "elasticity" of utility) is

$$\frac{du^*/u^*}{dI/I} = \frac{mx}{1-x} \left\{ \ln[(m-1)x + 1] - \ln\left(mx^{(1+\frac{1}{m})}\right) \right\} \tag{7}$$

The organization's utility function may be written as a weighted linear combination of the individual sub-group utilities as follows.

$$U = \sum_{i=1}^{n} z_i u_i^* \tag{8}$$

where the z_i's represent the weight attached to the importance of the ith sub-group's contribution and there are n sub-groups.

If P represents the price of a commodity and q the quantity of commodity consumed, the "revenues" from computing will be

$$P_1 \sum_{i=1}^{n} q_i = P_1 Q_1 \text{ (say)}$$

The cost of providing the computing service will be some function of the quantity of the computing commodity provided, say $f(Q_1)$. Thus the budget constraint of the organization is

$$\begin{matrix} \text{Fixed} \\ \text{resources} \end{matrix} + \begin{matrix} \text{Net income} \\ \text{from computing} \end{matrix} = \pounds + P_1 Q_1 - f(Q_1) \sum_{i=1}^{n} I_i \tag{9}$$

The necessary conditions for a maximum utility at the organization level, U, can be obtained from the equation

$$z_i \frac{du^*}{dI_i} = \text{constant} \left\{ 1 - \left(P_1 - \frac{\partial f}{\partial q_{i1}} \right) \frac{\partial q_{i1}}{\partial I_i} \right\} \tag{10}$$

186

Some explanation of this last equation is needed. The constant is a Lagrangian multiplier (i.e. a number), the left-hand side of the equation is the increase in the overall utility of the organization for each pound of budget allocated to user i, the term $(\partial f/\partial q)$ is the marginal cost to the central organization of providing an additional unit of computing power, $(p_1 - \partial f/\partial q)$ is the marginal contribution earned by the organization from the computer as a result of unit increase in quantity demanded at constant price, and $\{1 - (P_1 - (\partial f/\partial q_{i1})(\partial q_{i1}/\partial I_i)\}$ is the net amount by which the organization's resources are reduced for each marginal pound allocated, through I_i, to user sub-group i.

Since the constant multiplier is common to n such equations, eqn. (10) says that the optimal allocation of budget to leaders is one in which the ratio of marginal utility derived by the central organization from a pound allocated to use i to the corresponding net budget drain is the same for all leaders.

Smidt uses analysis further to solve the problem of optimal price setting and proves that the computer services should be priced at their marginal cost to the organization, i.e. $p_1 = \partial f/\partial Q_1$. Then the net budget drain is unity.

To obtain some feeling for the sensitivity of propensity to use the computer services to properly applied pricing policy, consider the case when the marginal propensity to consume computing services $(\partial q_{i1}/\partial I_i)$ is positive for all users. If the price of computer services is greater than their marginal cost then the net budget drain is less than unity. In this case the computer centre, at the margin, is profitable for the central organization. In these circumstances the organization, in maximizing its utility, would make larger budget allocations to users with a high propensity to spend money on computing than would be the case if marginal cost equals price.

In real organizations there would appear to be no uniform method of charging users for computing services provided by the central organization. However, it can be said, in general, that annual charges never exceed the running costs (i.e. they are a non-profit service) and in some instances management regard computing as a service to be provided free to those who wish to use it. In this case the user is constrained only by "turn-round" time (and if the load is high the computer manager must exercise some priority scheme) and his personal propensity to compute. High utilization, in theory, leads to low unit costs and if the demand is price-sensitive, the demand should further increase and vice versa. This suggests an instability in the price—demand relationship. Demand for computing does not appear to reduce under these conditions, presumably because the user is not sensitive to the price. He may not receive his bill until the end of the year and even though he does note its contents, he may be shielded from its significance by the central organization. Also the market is expanding in terms of users and would perhaps still continue to do so even if the costs were constant.

Usually there are two basic systems used by central organizations to provide computing capacity, a free service where demand is constrained only by propensity to use a computer and a charging system which encourages high utilization. There is no real problem with either system until such time as demand exceeds capacity at peak load, in which case queueing priorities are applied and/or higher prices are charged for peak load periods. Computing is a perishable commodity in so far that computing time not used is a wasted resource. Hence there is an incentive for central management to encourage full utilization. Nevertheless, utilization on its own is a poor index of utility because computers can be misused to provide "garbage-out from garbage-in".

Postulants for new computing services are often prompted by one of the following causes.

(1) Existing computers are not really capable of managing larger scale or more complex calculations.

(2) The current demand for computing time cannot be satisfied by the existing computers. Although the actual processing time may satisfy the needs of the sub-groups, in practise the total loading makes the average "turn-round" time excessive.

The second problem can be overcome by sharing, up to a certain limit, but the former may justify a larger, faster, or more complex computer. It is when requests for new computing services are made on the grounds of the need for more computing finesse that central management is faced with weighting subjective judgements and having to equate them in some way with real budgetary constraints. To reiterate, the work done by myself and Walford, and described here, was intended to develop a method whereby central management can assess, on grounds of cost—effectiveness, the need for new digital computing services.

To facilitate a less mathematical presentation of the analysis, the rest of the example will assume that the central organization has equated the marginal cost of computing to the price of computing. (The case where marginal cost does not equal price leads to

188

a viable analysis but is not appropriately included here because the algebra is already unavoidably "heavy".) With this last simplification, eqn. (10) becomes simply

$$z_i \frac{du^*}{dI_i} = \text{constant (which is the same for all sub-groups)}$$

$$= L \text{ (say)} \tag{11}$$

Thus the utility of the central organization is

$$U = \sum_{i=1}^{n} z_i u_i^* = L \sum_{i=1}^{n} u_i^* \Big/ \frac{du_i^*}{dI_i} \tag{12}$$

Since the *weightings* of the individual sub-group's utilities are assumed not to vary with funding in the short-term and a specific value of constant L yields specific values of (du_i^*/dI_i), it is convenient to normalize the value of the central organization's utility so that $U = 1$ when all the u_i^*'s equal 1. Thus

$$\sum_{i=1}^{n} z_i = 1 \tag{13}$$

and

$$L = 1 \Big/ \sum_{i=1}^{n} \frac{dI_i}{du_i^*} \tag{14}$$

Before carrying out any test calculations which approximate to real decision making needs, graphed versions of many of the above algebraic relationships were drawn with the help of a desk computer.

In the use of the above equations and/or their graphs and associated programme cards, some thought has to be given to the kind of input information required. Inputs to the "model" can comprise

(1) existing total funding and the allocation of funding between computing and non-computing commodities for each individual sub-unit,

(2) fixed and variable costs of existing and any other proposed computer system,

(3) and either (a) the present utility of each sub-group esti-

189

mated by central management and/or the leader of the sub-group or (b) a quantitatively ranked list of the relative importance of the contribution of each sub-group to the central organization or (c) a definition of the importance the central organization (or some higher authority) attaches to particular kinds of sub-group activity.

In algebraic forms these inputs would be

(1) All the values of S_1 and S_2 for every sub-group which are easily found from current accounts.

(2) The values of F (fixed cost) and v (variable cost) when the cost of computing approximates to the relationship [1]

$$f(Q_1) = (F + vQ_1)$$

(3) (a) all present values of $u_i{}^*$ or
    (b) all present values of $(z_i u_i{}^*)$ or
    (c) all imposed values of z_i.

Of the alternative input-data requirements, (3) (a), (b) and (c), the latter two are the easiest to obtain for psychological reasons. To obtain the values of u^* required in (3) (a) one would have to explain in some detail the concept of utility (and its consequence of diminishing returns) and perceptive leaders of sub-groups could possibly make self-interested calculations hopefully to ensure further increments of funding which are not so marginally small as to be almost unnoticeable (i.e. from the selfish point of view it may not be politic for a leader to say that his sub-group has achieved a high value of utility). However, the requirements (3) (b) or (3) (c) may be obtained without explaining all the intricacies of utility functions.

When the current apportionments of spend, (1), are all known together with *one* of the measures of achieved, or desired, sub-group "value", (3), the intrinsic parameters (k_1 and k_2) which are indicative of the usefulness of computing and non-computing activities, may be calculated.

The computational procedure for obtaining values of k_1 and k_2 are straightforward (but boring) and space-consuming to illustrate. As an indication of procedure, the calculation is outlined for the

---

[1] Obviously it would be highly desirable to have the complete cost function for computing (i.e. $f(Q_1)$) but in most cases the best one can hope for is a straight-line function which approximates to this function over a small range in Q_1.

case where the central organization has quantified proportional contributions currently made by sub-groups (i.e. the values of $z_i u_i^*$) in abbreviated form using the following flow diagram.

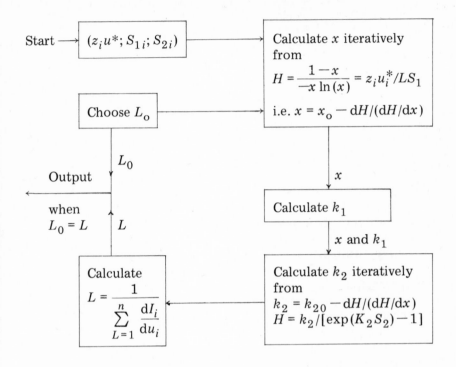

Using this procedure Medford and Walford considered four *fictitious* sub-groups, with widely different apportionments of spending, as indicated below.

Sub-group	I_i	S_1	S_2	$Z_i u_i^*$
A	£300,000	£100,000	£200,000	0.17(5)
B	£110,000	£ 10,000	£100,000	0.09(2)
C	£ 60,000	£ 50,000	£ 10,000	0.07(2)
D	£260,000	£250,000	£ 10,000	0.22(8)

The outcome of the calculation was a convergent solution which gave the following parameters.

Sub-group	u^*	K_1	K_2	$\dfrac{du^*}{dI}$	z	L
A	0.5	0.02	0.0043	0.00157	0.35	
B	0.4	0.396	0.0052	0.0030	0.18	(1/1,820)
C	0.8	0.034	0.400	0.0061	0.09	
D	0.6	0.0037	0.541	0.0015	0.38	
					$1.00 = \sum\limits_{i=1}^{4} z_i$	

A change in the overall level of funding will change the value of the constant (L), $(S_1, S_2, x, u^*)i$ etc. but it will not change the intrinsic values k_{1i}, k_{2i}, and z_i. (Remember that the k's are utility co-efficients determined by the nature of the sub-group's research activity and the z_i's are the weightings attached by the central organization to these activities.)

To find out how funding decisions change the individual and sub-group utilities, it is only necessary to vary L progressively upwards and downwards and obtain the corresponding values of the parameters S_{1i}, S_{2i}, I_i, u_i^*, and U. For the example in hand, a result of this procedure is shown as a curve in Fig. 2 which depicts the change in central organization's utility and internal allocations, within sub-groups, between each commodity.

Naturally, once the parameters that represent the systems have been derived, this cost—effectiveness model is in a suitable state to

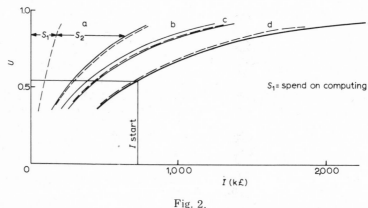

Fig. 2.

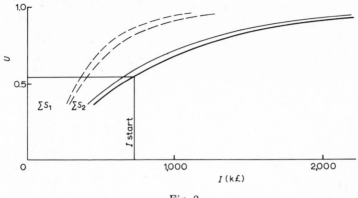

Fig. 3.

assist decision making on new technology. This is demonstrated below.

To find the Change of Central Organization's Utility (U), and other Parameters, consequent upon a change of Computer Technology

New computer systems generally yield a lower cost of processing a work load. This really means that k_1 is increased as computer technology becomes more advanced. If the central organization's utility is to be kept optimal, computing price must be dropped commensurately with reduction in computing cost.

Suppose the results derived from the model, and shown in the last two tables, are taken as a start point and the value of k_1 is enhanced by 21% as a result of improvements in computing technology. Then the use of the model (not given in detail here to avoid mental indigestion through unnecessary detail) yields Fig. 3 which shows, at a glance, how the overall utility is increased for the same sum of all allocated budgets. But the real problem in introducing a new computer system is "is it worth while?".

To obtain an answer one must return to the budget constraint on the central organization, namely

$$\pounds + \left(1 - \frac{v}{P_1}\right) \sum_{i=1}^{n} S_{1i} - F = \sum_{i=1}^{n} I_i \qquad (15)$$

where F is the fixed cost of computing, v is the variable cost of

193

computing, and P_1 is the price of computing. Then if it is assumed that variable cost is again put equal to price in the new system to obtain optimal conditions and subscripts o and N represent old and new computing systems respectively, the conditions after and before the installation of the new system may be represented as

$$\pounds_N = F_N + \left(\sum_{i=1}^{n} I_i \right)_N - R$$

$$\pounds_o = F_o + \left(\sum_{i=1}^{n} I_i \right)_o$$

(16)

where R denotes the resale value of the old computing system. If it is a requirement that the new central funds shall equal the old central funds the increment, or decrement, of funding made available to the sub-groups is given by

$$\Delta(I_{total}) = \left(\sum_{i=1}^{n} I_i \right)_N - \left(\sum_{i=1}^{n} I_i \right)_o = R - F_N + F_o$$

(17)

Usually the fixed cost of the new computing system is larger than the fixed cost of the old computing system and if the resale cost of the old computer is not large, the funds available to the sub-groups as a result of the installation of the new system can be reduced. The decision in this case depends upon whether the enhancement from the increment in computer efficiency compensates for the necessary reduction in funding (which need be no hardship if the price of computing has been reduced). In terms of the graph in Fig. 3 this depends on whether the horizontal shift between the curves is greater or less than the magnitude of $R - F_N + F_o$. (Note that the overall utility curves for the old and the new system are, in this fictious case, very nearly parallel.) When central funds are to be enhanced to encourage the adoption of the newer computing system (i.e. $\pounds_N - \pounds_o = \Delta \pounds$ is greater than 0) the expected change in overall utility, U, is obtained by moving onto the new curve with a shift of $\Delta \pounds + R - F_N + F_o$. Sometimes the result will be that central utility is enhanced, sometimes it will be reduced. When the central utility is reduced, quite obviously the proposed system is not worthwhile; when it is increased the arbiter-of-decision has got to decide whether the increase is worth the greater expenditure.

194

Concluding Comments

Intentionally, a rather complicated example of cost—effectiveness technique has been given above because in technology assessment the difficult problems, of assessing the impact of technology on social and economic environments, are hardly likely to be resolved without some resource to the use of utility functions. Although the given example is seriously meant, it is not now appropriate to branch off into a specific discussion of the difficulties which arise from the application of the model. Questions such as: "Are computer systems operated in a free-market?"; "Is utility the appropriate parameter to consider in cost—effectiveness analysis of computer systems?"; "If it is does the form of utility function described above adequately cover the situation?"; and "Is it true that leaders can, and do within their limitations, optimally allocate the different resources at their disposal?" represent major queries which users of the model should address themselves to.

Less ponderous, and perhaps less meaningful, examples of straightforward cost—effectiveness analysis are easy for the reader to conjure up for himself. Very simply, such examples will involve sensitivity analysis, like that described in the previous chapter, but with a quotient which is comprised of a numerator which is some parametric representation of effectiveness and a denominator comprised of cost.

A bachelor, for humorous example, in pursuing his amorous adventures might rate success with a girl as providing satisfaction S (definitely unquantifiable). In addition, depending upon his natural charms, he will incur some sort of cost function in pursuing the fair sex. Suppose the cost function consists of fixed and variable costs (accomodation and plumage, average cost per entertainment) such that the total cost of courting n girls is $£(F + vn)$ and the probability of success with any girl is, on average, equal to p (which is less than unity). Then the persistent bachelor's satisfaction per unit cost, expressed in terms of the above parameters, is

$$\text{Effectiveness} = \frac{S(\text{expectance of successful wooing})}{(\text{cost of attempted wooings})}$$

$$= \frac{S.np.(1-p)^{(n-1)}}{F + vn}$$

Graphical or analytical optimization of this quotient will reveal

195

that the optimum value is achieved when the bachelor[1] woos

$$\frac{+ F\ln(1-p) + [(F\ln(1-p))^2 + 4vF(-)\ln(1-p)]^{\frac{1}{2}}}{2v(-)\ln(1-p)}$$

girls, which fortunately for the bachelor, is always a positive number since the logarithm of a number less than unity is negative!

Reference

Smidt, S. (1968). "The Use of Hard and Soft Money Budgets" *Proc. Conference AFIPS*, Vol. 33, Pt. 1.

---

[1] Amathematical bachelors should either use graphical methods to obtain the optimal number or proceed confidently in the non-cost—effective way. Married men are warned that the numerator should, in their case, contain a negative term to represent the expectance, and consequent anguish, of being found out.

CHAPTER 7

Technological Forecasting and its Place in Technology Assessment

"A man must conform himself to Nature's laws, be verily in communion with Nature and the truth of things, or Nature will answer him, No, not at all! Speciosities are specious — ah me! — a Cagliostro, many Cagliostros, prominent world-leaders, do prosper by their quackery, for a day. It is like a forged bank-note; they get it passed out of their worthless hands; others, not they, have to smart for it. Nature bursts-up in fire-flames. French Revolutions and suchlike, proclaiming with terrible veracity that forged notes are forged."

Thomas Carlyle (1840)

Introduction

In a sense technological forecasting fathered technology assessment. Participation in the act of forecasting potential technology led many scientists and technologists to consider more carefully than had been their wont, the rationale behind the setting up of social goals. This, in turn, focussed more attention upon the real need to manage technology more effectively in order to achieve societal goals. Before the popularity, said by some to be excessive, of technological forecasting, there was undoubtedly much incoherent talk of social goals by the scientific fraternity. After sermonizing on technological forecasting, however, discussion on social goals became somewhat more coherent because the wide, all-

197

embracing view demanded of the forecaster led the "hard" scientists to ponder subjects that they had previously thought to be only appropriate to the "softer" social sciences. The basic techniques of technological forecasting are, therefore, of historic as well as practical significance to the student of technology assessment.

This chapter gives sufficient introductory material on technological forecasting to allow the reader to make up his own mind about the value of the pot-pourri of techniques which comprise the subject and gain an appreciation of its place in technology assessment. It is possible that this selection of techniques will exclude some favourite methods because a complete anthology requires a book to itself. Nevertheless, the reader who wants to study technological forecasting exhaustively has already got many excellent specialized books at his disposal, for example Jantsch (1967) and Cetron and Ralph (1971).

Exploratory Techniques of Technological Forecasting

Exploratory techniques comprise methods of extrapolating present trends, indicated by past and present data, into the near and far futures. Such extrapolations can, naturally, be made with or without the aid of mathematical techniques but, in general, mathematical techniques are used and can be categorized, as

(1) techniques for ensuring the best statistical fit to the known data before extrapolating the curve into the unknown, and

(2) techniques for fitting current data to a curve whose general shape is predetermined by some known, or implied, causal relationship.

As an illustrative example of extrapolation without the aid of mathematical analysis, we might consider an early PAU evaluation of a desalination programme (Medford, 1969) where the technology under consideration was aimed at satisfying the potential demand for fresh-water between 1980 and 2000.

The problem was to forecast the need for technological innovation and for this purpose the PAU used a study carried out by the Water Resources Board which showed that by the end of this century the excess of demand, in S.E. England, over 1965 supply would amount to 1100 million gallons per day (unbroken line in Fig. 1). A number of schemes in hand or proposed at the time of the analysis (see Table 1) were to augment supplies to meet esti-

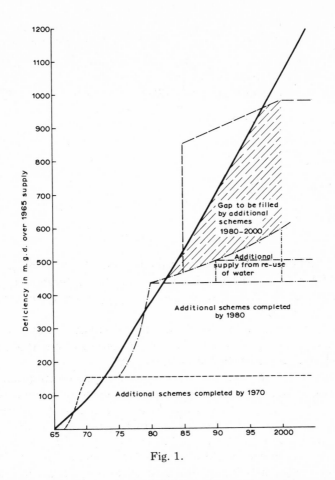

Fig. 1.

mated demands up to about 1980, and these have been superimposed on Fig. 1. To meet the excess potential demand between 1980 and 2000 various alternative schemes were suggested.

(*1*) A barrage across the Wash estuary which would supply about 380 million gallons a day. Supplies of fresh water would not become available from it, however, until the completion of the whole scheme (which would take about 15 years). This sudden large influx of additional supply would inevitably result in a period of years when demand exceeded supply or vice versa, depending upon the time of its introduction.

(*2*) The introduction of a number of smaller schemes from 1980 onwards such as diversion of water supplies from other parts of the country, the artificial recharge of aquifiers or the provision

TABLE 1
Proposed additional water schemes for south east England

Scheme	Total increase (m.g.d.)	Total capital expenditure (£ millions)	Phased expenditure from 1966			
			66/70	71/80	81/90	91/00
Authorities	130	19.5	17	2.5		
Local schemes	15	16	8	8		
Diddington	15	4.7	4.7			
Ely Ouse	22	8.5	8.5			
Great Ouse	90	29.0	1.0	28.0		
Thames G.W.	125	14.0	3.4	10.6		
Welland and Neve	20	7.0		7.0		
Re-use	215	15.6	0.7	3.3	5.1	6.5

of desalination plants. The commissioning of desalination plants of 100–150 million gallons per day at four year intervals from 1985 onwards would provide increases in supply broadly matching increases in demand.

(3) Increasing the utilization of recirculating water by new methods of purifying water. This could help to bridge the gap but, in itself, would probably not be sufficient.

This illustrative example of non-mathematical extrapolation is, of course, much simplified. It takes no account of relative economics nor offers any solution to a complex problem. It does, however, help to define boundaries of the problem, identifies possible solutions, and, by extrapolating the gap to be filled by additional schemes between 1980 and 2000 (the shaded area in Fig. 1), indicates future deficiency which there were no plans to meet. In this case recognition of the very important part re-use of water had to play drew attention to the possible application of desalination techniques to sewage treatment.

The example might fairly be criticized because, although it is indeed an example of extrapolation without mathematical technique, it depended essentially upon the use of a previous forecast made by the Water Resources Board of excess demand which could itself have depended upon mathematical technique. We now deal with mathematical techniques of extrapolation.

For mathematical methods of determining the best curve fit to the current data prior to extrapolation, the reader should refer to standard text books on the elementary theory of statistics. Straightforward curve-fitting techniques cannot rightly be claimed to be more than a statistical aid which just happens to be applied in technological forecasting. In contrast, the selection of causal relationships and methods of fitting them to the available data can be considered an intrinsic part of technological forecasting. Undoubtedly the most common of these causal relationships (and perhaps the most overused) is the biological growth model alias, the logistic curve, the S curve, and the sigmoid curve.

Pearl (1924) discovered that the following formula gave a good representation of biological growth phenomena (e.g. increase of fruit flies within a bottle, cell increase in white rats).

Number at time $t =$

$$= \frac{\text{Upper asymptotic limit}}{1 + [\text{Constant} \times \text{Exponential of } (t \times \text{a negative constant})]} \quad (1)$$

This S curve formula has a graphical appearance somwhat as shown in Fig. 2.

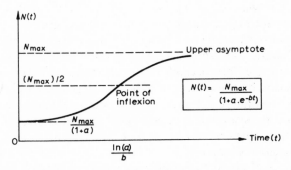

Fig. 2. The 'S' curve

From the above, the only "causality" obviously apparent is that no biological system can grow to an indefinite size and there must be an upper bound to growth, hence the asymptotic limit and the relatively slow growth as this limit is approached. However, the above empirical formula does yield a very interesting expression for the *rate* of growth,

$$\text{Rate of growth} = \frac{dN(t)}{dt} = \text{constant} \times N(t) \times (N_{\max} - N(t)) \quad (2)$$

which means that the rate of growth is proportional to present size *and* the amount of growth still possible before the ultimate size is reached, a causal relationship which is intuitively plausible.

The same causality can be inferred in other situations not directly related to biological growth, for example consider the psychological mass phenomenon of acquisition of goods, or attributes, where the population is uniformly dispersed. Those people who already possess the good, or attribute, transmit their satisfaction to those who do not possess it and the rate of possession will, all other things being equal, grow proportionally to those who have it multiplied by those who do not. This is simply a neat mathematical way of expressing the phenomenon of psychological pressure upon the non-owner.

Lewandowski (1972) gives many examples, which are well worth study, of the use of the logistic function to make straightforward forecasts of market demand. Similarly, Fisher and Pry (1971) give statistically well documented evidence for the plausibility of the logistic model in studying the substitution of one technological good for another, synthetic for natural fibres, detergents for natural soaps, organic for inorganic insecticides, etc.. In the Fisher and Pry example, the variable determined by the logistic function is the fractional substitution of new for old and eqn. (2) fits existing data so well, in 17 cases, that the authors recommend its use for forecasting technological opportunities, recognizing the onset of technologically-based catastrophes, investigating the similarities and differences in innovative change in various economic sectors, studying the rate of technical change in different countries and different cultures, and investigating the features limiting technological change.

Dangers of S curve forecasting are that the analyst can never be absolutely confident of the values of the constants in eqn. (1) and, indeed, not even sure that the constants do not vary with time. Most growth curves which span long intervals of time are not particularly smooth and present a jerky appearance. Possible explanations are that the overall growth is composed of superimposed S curves representing successively better innovations or straightforward replacements (see Fig. 3) or, even worse, that the underlying causality implicitly assumed when using a logistic function, is wrong.

In regard to the last statement, there is an almost classic case of the partial verification of causality by Coleman *et al.* (1957) which is also reported in Coleman (1964). It appears that in investigating the adoption of a new drug by American doctors, the analysts

202

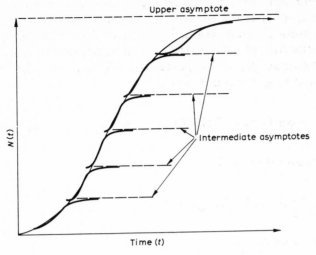

Fig. 3.

could not obtain a good fit to the data by using, separately, either a logistic function or an exponential growth function of a kind appropriate for single impetus to growth[1]. However, when the total population of doctors was separated into two categories, the integrated and the isolated doctors, that is doctors who had the desire and the opportunity to be gregarious and those that had not, good statistical fits to two causal relationships could be obtained. The gregarious doctors seemed to obey the logistic function in their adoption of the new drug, and the non-gregarious the exponential function. This is precisely what one would have assumed from consideration of causality. One would expect the gregarious doctors to be amenable to persuasion to adopt the drug through social contact with other doctors who had already adopt-

---

[1] The exponential growth model for a limited population N_{max} rest upon the causality that the number of people who acquire the attribute, or good, as a result of a single source of persuasion is proportional to the number that have not made the acquisition, i.e.

$$\frac{dN(t)}{dt} = \text{constant} \times (N_{max} - N(t))$$

When integrated, the above rate-of-growth formula yields

$$N(t) = N_{max}[1 - \exp(-\text{constant} \times t)]$$

a form which, when plotted, is concave towards the time axis and never exceeds the asymptote N_{max}.

ed it and also through advertising persuasion. Conversely, one would expect that the isolated doctors would be deprived of persuasion through social contact and have a rate of adoption, for a given intensity of advertising, which was proportional only to those who had yet to be persuaded. We can state this result in mathematical symbols as follows.

$$\frac{dN(t)}{dt} = \text{constant} \times N(t) \times (N_{max} - N(t))$$

for gregarious users and

$$\frac{dN(t)}{dt} = \text{constant} \times (N_{max} - N(t))$$

for isolated users.

Quick perusal of the shape of the simple S curve, or logistic function, sketched in Fig. 2 will show that, whatever the values of the three constants which determine the shape of the curve, the point of inflexion is always at half the asymptotic value. It is not expected, of course, that such neat "symmetry" will always pertain in a real life situation even though the causality underlying the functional form of the curve is plausible. Accordingly, attempts have been made to generalize the logistic function so that the symmetry is not inherent. A whole set of generalized logistic functions is used by forecasters and these are better studied in the specialized literature, see for example Lewandowski (1972). For the record, the general form of the most common of these functions is recorded below.

Generalized logistic function of the 1st order

$$\frac{dN(t)}{dt} = \text{constant} \times N(t) \times [N_{max} - N(t)]^{\text{constant}}$$

Generalized logistic function of the 2nd order

$$\frac{dN(t)}{dt} = \text{constant} \times [\ln (N_{max}/N(t)]^{\text{constant}}$$

Logistic functions of the first order are normally used when the point of inflexion is reached after the growth has exceeded 50% of the asymptotic value, i.e. a high growth rate during the starting phase and a slower rate during the second phase.

Conversely, logistic functions of the second order have a point of inflexion before the growth has achieved 50% of the asymptotic value and are characterized by a faster growth rate in the second phase. Another popular function which can be asymmetrical and represent growth tending to an asymptotic value is the Gompertz Growth Function which is represented by the following mathematical function:

Gompertz function

$$\frac{dN(t)}{dt} = \text{constant} \times [\ln(N_{max}) - \ln(N(t))]$$

Although all the extrapolation functions have served the technological forecasters well, it is remarkable that none of the functions popularly in use allow explicitly for assimilation delay. Perhaps the next phase in extrapolation modelling technique will involve the development of "delayed-growth" models, which make allowance for communication and other delays. To illustrate what is meant by this suggestion, consider a simple extrapolation model of "keeping up with the Jones's". The usual assumption is that the rate of acquisition of new goods or attributes is proportional to what the Jones possess, i.e.

$$\frac{dN(t)}{dt} \propto N(t)$$

which gives an undelayed growth of form

$$N(t) = N(t=0) \times e^{\text{constant} \times t}$$

However, one would expect the rate of acquisition to be proportional to what the Jones's are *known* to possess and it takes some time to discover that Jones has got in first again. Therefore, the causal relationship should look like

$$\frac{dN(t)}{dt} \propto N(t-T)$$

where T is the average time to discover that Jones has already acquired the good or attribute.

One solution of the last equation looks, but is not, complicated.

$$N(t) = N(t=0) \sum_{m=0}^{\infty} (\text{constant})^m \times (t-mT)^m \times U(t-mT)$$

where

$$U(t-mt) = 1, \, mT > t$$
$$= 0, \, mT < t$$

The form of this delayed growth equation is sketched in Fig. 4, and it will be seen that as the delay tends to zero the growth becomes equal to the exponential growth of the conventional model.

Other more complicated delayed growth models are known (see Medford (1970) for the delayed Gompertz Curve) but these will not be described in this book. Indeed, the only reason for giving the previous detail is so that the effect of "assimilation" delay on extrapolatory forecasting can be demonstrated.

Extrapolatory forecasting by analysis of precursive events is recognized as an acceptable technique when there is a causal relationship between the known, earlier development and the expected future development. Lenz, an esteemed pioneer of technological forecasting, gives a good example of this in which the precursor is the maximum speed of military aircraft and the development to be forecast is the upper speed of civil transport aircraft (see Lenz, 1962). If we consider, however, a completely new field of human endeavour without any obvious precursor, are there any situations in which this method of trend correlation is applicable?

A perhaps lighthearted, and certainly speculative, exercise

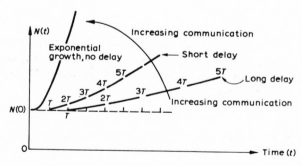

Fig. 4. A simple delayed growth model.

206

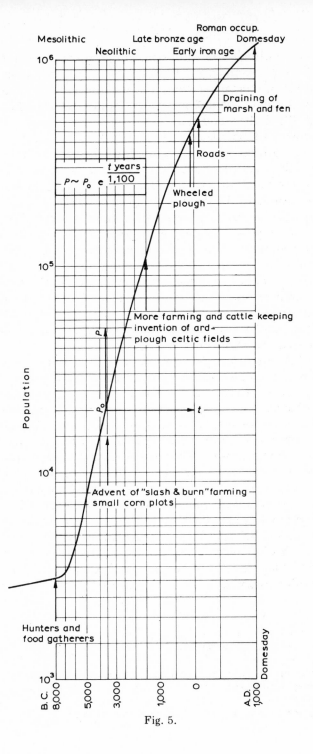

Fig. 5.

207

would be to estimate the future growth of human undersea population. A comparison might be drawn with the development of land cultivation beginning with the neolithic period when "slash and burn" farming enabled our ancestors to extend their activities from the hills to the wood valleys.

The growth of population (see Wood, 1967) is plotted against time in Fig. 5 from which the growth of the agricultural population, from the end of the mesolithic period to Doomsday, is found to be exponential with an e-folding time of 1,100 years[1]. Can anything meaningful be deduced from this about the development of the submarine environment? Even though pioneer aquiculturalists will be equipped with technological and medical aids and means of communication and environmental control which were unthought of even fifty years ago, they will face hazards and possible catastrophe no less severe than their pre-Doomsday agricultural forebears. The differences in kind, however, may be regarded as too fundamental for an analogy of this kind to be valid but at least we have a starting point for debate and a means of concentrating attention on some of the factors which seem relevant. Rather provocatively, therefore, it is possible to estimate that the rate of population growth of human aquiculturalists will not exceed about 250% per millennium.

The real skill in extrapolatory forecasting lies not in the discriminating use of mathematical techniques, though this is a very necessary accomplishment which somebody in a technology assessment team should possess. The finesse lies in the selection of the parameter, or parameters, to be extrapolated. To give a trite example: to obtain a useful index of future digital computer performance, a single parameter may not suffice but a combination of one or more parameters may be more meaningful particularly when we presently have difficulty in perceiving the details of future technology. In this last regard, the American forecaster Robert Ayres in his book *Envelope Curve Forecasting*, has shown that a particularly apposite combination is storage capacity divided by the time

---

[1] For the non-mathematical reader, the term "exponential growth" requires some explanation. It simply means growth that increases itself e times (e = 2.718) after every time lapse equal to the e-folding time. Thus in the above use, one would expect the undersea population to multiply by 2.718 times every 1,100 years. Exponential growth may be regarded as growth according to the rules of compound interest see Rose and Rose (1971) for cogent comments on the exponential growth of science.

taken to make an addition, *viz*[1].

$$\frac{\text{Storage capacity in bits}}{\text{Add time in seconds}} = 10^{7+0.39(t-1948)}$$

The Ayres' formula is really a proxy measure of the tendency of computers to become increasingly faster and capable of performing tasks of increasing difficulty.

Analysts of industrial markets have known for years the need to use proxy measures of the things that should, ideally, be measured but do not lend themselves to direct quantification[2]. Goodman (1972) discusses the point from the marketing point-of-view and gives the following example, of the use of proxies, taken from Spencer and Siegelman (1964).

Demand for machine tools = Function(X_1, X_2, X_3)

where X_1 is an index of metal and metal products weighted by average labour wage rates. This is a proxy for the total amount of service to be received from machine tools and the labour savings attainable through new machine purchase. X_2 is the percentage of its capacity presently being used by the metal products industry. This is a measure of the user's existing stock of problem-solving capacity. And X_3 is the profit index of metals and metals product industry. It is a proxy of the industry's ability to pay for new machines.

Unfortunately, in these days of highly paid consultants, "accurate" extrapolatory formulas are not easy to obtain without making payment. A complete compendium of successful formulas is, therefore, beyond the means of this author. Nevertheless, formulas for extrapolation, using either real of proxy data, do appear in the literature and the best advice I can give the aspiring technological assessor is to make a personal collection of those formulas that are appropriate to the nature of his business. Associations of

---

[1] This formula is just another arithmetical form of exponential growth. It is not surprising that many extrapolations yield exponentials, and combinations of exponentials, because Mother Nature seems to favour the exponential laws of growth and decay.

[2] Market analysts are really a specialized sub-group of technological forecasters. Their efforts to predict and possibly affect demands for specific goods are necessary part of obtaining "a probabilistic assessment on a relatively high confidence level, of future technological transfer".

technological forecasters make ideal exchange marts for swapping this kind of information and an appropriate organization, for Europeans, is the European Technological Forecasting Division of the European Industrial Marketing Research Association (EVAF, London).

Normative Techniques of Technological Forecasting

Normative forecasting attempts to assess future goals, needs, desires, missions etc. and then works backwards to the present identifying what changes are required to take us from the present state to the normative goal. Whereas there is something fatalistic about exploratory forecasting because it suggests that the present and past inexorably determine the future, there is something optimistically deterministic about normative forecasting. Intrinsically, there is a self-fulfilling aspect to prophesies based upon normative forecasts, simply because such forecasts are usually used to set mission-orientated goals. For obvious reasons, normative methods are favoured by organizations which are large enough to proceed along the normative path mainly under their own volition. Thus the technique is commonly used by government departments, very large companies, and organizations that have, or think they will obtain, political power. Within a limited range of freedom, smaller and less powerful organizations can plan on the basis of normative forecasts but such planning has so large an exploratory content, and so little freedom to choose desired goals without outside influence, that the designation normative forecasting is hardly appropriate.

Generally, quantitative normative forecasting tends to depend upon what have become known as Relevance Techniques which are mathematical methods of arithmetically ranking the alternative paths possible from the present to the designated normative goal. Eavesdroppers of war-time conversations will recollect that the commonly advocated solution was to dispose of all enemies by pressing the activator of some gargantuan weapon of destruction designed for genocide[1]. The same listeners, overhearing recent discussion on relevance numbers and technological forecasting could

---

[1] These conversations were noticeably muted after the first atomic bombs were used and the conceptual solution, borne out of frustation and fright, became a reality.

possibly be forgiven for assuming that some analysts, bemused by the absence of fact in complex forecasting situations, were proposing analogous solutions, only this time the worries are to be removed by the use of cabalistic numbers.

Basically, my own worries about relevance techniques do not focus on the use of numbers (even when they are on an ordinal scale and it is impossible to describe the relative difference between alternative approaches scoring, say, 2 and 3 as opposed to 12 and 13). They focus upon the almost irreversible nature of the process as viewed by the final policy maker when he is presented with the ranked alternatives in descending order of magnitude. The final arbiter-of-decision may not be able to determine quickly how the number was arrived at. He may know the mathematical processes used by the assessors or forecasters in deriving the number and be aware of any inherent logical inconsistencies, but a number may have been obtained in different ways without varying the underlying processes.

In any relevance system there is, of necessity, some quantification of subjective opinions; that is, experts will be asked, say, to attribute a number between 1 and 10 to the usefulness of a segment connecting two well-defined stages in the normative path. The quantified subjective opinions are then fed into the mathematical process which determines the overall "relevance" score of the complete path from a given starting point to the normative goal. Such arithmetic ranging of subjective opinion may be a useful expedient in decision making and forecasting. If it is, then the following, previously used, quotation should give salutary warning to readers before they move on to the following detailed consideration of relevance techniques.

"This process leads to decision outcomes that have such clarity and force that they can withstand even the strongest questioning, as Congress was taught in the controversial TFX investigations. This situation illustrates the hazards produced when the lack of facts leads to overcompensation in elaborate techniques for handling non-facts".

Edward B. Roberts

A European Relevance Technique

Monsieur H. de l'Estoile, Directeur des Affaires Internationales, Délegation Ministérielle pour l'Armement, Ministéres des Armées, France, whilst working in the Centre de Prospective et d'Evaluations (CPE), outlined a pioneering relevance technique used by the military. John Hobbs, acting as a consultant to the Programmes Analysis Unit in 1967/68, gave a factual exposition of de l'Estoile's 82 pages of French text. The following notes are culled from Hobbs' account and are given here because readable English versions of de l'Estoile's technique are not readily available.

The CPE use the relevance technique to evaluate prospects for military R & D projects but de l'Estoile also described how the system could be adapted to aid planning in the field of civil transportation instead of defence.

Members of the French consulting company SEMA (Societé d'Economie et de Mathematique Apliquées) advised the CPE on the internal logic of the method, but the basic initiative came from the CPE. The method was used by SEMA to evaluate their own research programmes and the CPE is reportedly still continuing to improve the method.

Topics covered in this account are

the general background to the problem of decision making in CPE,

general description of the method,

comments on the method, and

applications to the Civil sector.

Annexes to this account are concerned with

the rules for combining subjective rankings and

the economic cost—risk factors.

It is not simply a masochistic exercise for the author to regurgitate details of de l'Estoile's (and later on an American) relevance technique. I put myself to the task of explanation in order to indicate how details thrown up by normative forecasting exercises led to greater concern about methods used for selecting normative goals (particularly the social ones) and to illustrate the horrendous task that still awaits assessors of technology and their colleagues in the social sciences before they can give structured advice upon the social impact of technology. As with the criticism of the Byatt—Cohen model in Chapter 8 for the study of benefits likely to stem from curiosity motivated research, some hard things are said below about relevance techniques. Nevertheless, as with the Byatt and Cohen work, I have much pleasure in acknowledging the courage of the French and American analysts who did the pioneering work on relevance techniques and repeat the rider, from Chapter 8, that critics like myself are often the first beneficiaries of their work.

General Background

The relevance method of de l'Estoile is used by the CPE in the following context.

Some 800 to 1,000 research proposals have at any one time to be reviewed, assessed, and placed within a technological forecast and plan.

The CPE analysts must make military relevance and utility their primary consideration. But if they can, they should denote any spin-off from projects to the civilian sector, particularly when these might sway the decision one way or the other for marginally attractive projects.

Cost and use of resources available to carry out projects are of importance. In the initial use of the relevance method, however, many organizations were available to carry out the projects and no early attempt to introduce specific constraints on cost and resources was included in the analysis.

In 1967, the time horizon for implementation of final outcomes of projects was between 1975 and 1980.

A group of analysts (defined by the French as "engineers") is available to the CPE for the implementation of the relevance system. These analysts have access to military experts for allocating importance, through subjective scores, to military strategies and tactics. They also have access to scientists and technologists for the assessment and evaluation of research projects.

The CPE acknowledge the low precision which can be expected from the quantitative methods used in the relevance analysis. Therefore, they seek to classify projects examined into three broad categories.

(1) Those which are definitely acceptable,

(2) those to be rejected, and

(3) the remainder.

The evaluation, using the relevance technique, is meant to be annual. When de l'Estoile reported upon the outcome of using the method in 1966 to 1967 there was obvious satisfaction with the conclusions it gave, although there was also a realization that the method stood in need of some improvement and continuing efforts to refine the methodology are expected.

General Description

The method assesses the military significance of projects against a hierarchy of five reducing levels of importance.

213

Hierarchical levels	Number of entities on each level
Goals dependent on Government defence policy	3
General missions or fields of activity	40
Particular needs or systems to help achieve missions	500
Sub-systems or technological capabilities to achieve the systems	1000
Elementary research projects which contribute to one, or more, of the desired technological capabilities (listed as "OUR" by the French)	800

A network, representing interconnections between the hierarchical levels, permits the military significance of an entity on any level to be described by attaching a score (rank number) to the importance of the particular element to the elements at the superior levels of importance. In the nomenclature of technology assessment, these networks have become known as "relevance trees" and a typical example is sketched below in the description of the American relevance system PATTERN.

The ranking scores permitted in the CPE relevance system are

0 indispensable

1 very useful

2 useful

3 useful in some circumstances

4 of very little use [1]

where this ranking describes the need for a particular mission, say the ability to transport men and their equipment 1,000 miles within 24 hours.

Combination of relevance ranking with ranking of the degree of need is achieved by defined combinatorial rules of which an example is given below. Ranking of relevance of a specific system to fulfilling the mission is measured by an index

0 absolutely vital to the achievement of the mission.

---

[1] I know not why the CPE analysts reverse the normal procedure of attaching higher scores to greatest usefulness. Those assessed might suspiciously assume that, when the inner workings of relevance systems are obscured, the reasons are psychological.

214

1 will considerably assist attainment,

2 might eventually assist attainment, and

∞ will play no part in attaining.

Such a system might be aircraft capable of carrying 100 tons at 600 knots. Combined ranking of the system which considers both its relevance and the need for the mission to which it is relevant, is represented by the appropriate number in the following matrix.

		\textit{Need for mission}				
		0	1	2	3	4
	0	0	1	2	3	4
Relevance	1	1	2	3	4	5
of system	2	2	3	4	5	6
	∞	∞	∞	∞	∞	∞

N.B. The combinatorial rule is simply additive and it is assumed that the combination (0,1) of need and relevance is the same as the combination (1,0).

The whole subject of the construction of these matrices is discussed in Annex A, since it is vital to the de l'Estoile relevance exercise. By classifying projects, sub-systems, and missions in a manner to that just illustrated and using elementary combinatorial rules which take into account attitudes about the relative importance of the factors or attributes used in the classification, a combined rank number (between 0 and 11) is deduced by the CPE for the military utility of a project.

In the French military context the index of military utility (U, say) is now combined with two other indices, the first depending on cost and risk (labelled q in the text) and the second depending on the economic significance of the project (labelled E in the text). Factors q and E are both described in Annex B. The overall ranking of the project combines first U and q and then combines $(U + q)$ with E to give a final rank number between 0 and 14. The military can, of course, ignore the economic significance E and clearly the degree of weighting they attach to it is mainly a matter of policy decision.

It is important to realise that all ranking numbers used imply an *ordinal* scale (it is impossible to describe the relative difference between projects scoring 5 and 6 as opposed to 11 and 12). Indeed, such is the accuracy of assessment that it is unlikely to be possible to say confidently that there is a significant difference between scores one integer apart. It is, however, claimed that projects at the low end of the relevance rank table are almost certain to be the ones that should be rejected first. In the CPE system outcome, the French rejected 20—30% of the projects examined [1].

Comments on the Method

(a) A complete network of the connections between levels in the hierarchy was never constructed. For each project its connections were indicated and the connections of elements at other levels were also listed in a not too systematic fashion. This needs improving, and will be tackled in the future.

(b) Multiple links between one item and several items in a superior level were only allowed for at the level of systems which were relevant to several missions. A rule was developed which, broadly speaking, says that if one ranking of relevance is much more important than others, then it remains the overall ranking. If, however, there are two or three virtually equal rankings then, under certain circumstances, the overall effect is to give a rank which is one unit more favourable. In the civilian applications it may prove worthwhile to calculate the net effect of multiple relevance at lower levels of the hierarchy but in the interests of simplicity the idea may be put aside in the initial stages.

(c) The relative values of the three separate goals were not established. This amounted to ducking the issue and relative values started at the mission level. In the civilian context, this amounts to a refusal to weight the relative importance attaching to say defence versus economic standard of living versus social conditions versus international prestige. The refusal to quantify is understandable.

(d) A computer is now used, but up to now (1967) CPE have managed with a somewhat unorganized manual system, mainly possible owing to their rejection of multiple linkages between items at the lower levels in the hierarchy.

(e) With regard to the difficulty analysts experience in deciding which word or item in a relevance tree corresponds to which level in the hierarchy, the CPE apparently let themselves spawn intermediate levels and inserted words or activities which had some

---

[1] More literary minded readers might prefer the other method of ranking decisions described in the amusing French novel *Clochmerle* where officials used a dartboard.

semantic linkage with other activities and tended to make it easier to visualize the logic of the inter-connections. Subsequently, one can ignore the intermediate levels generated between the officially recognized levels, merely using them as a guide in doing the ranking.

On the other hand, if it is logically necessary and useful to create another hierarchic level, then there is no objection to doing so provided the resulting rank number scale is consistent with that devised for the normal number of levels elsewhere.

(f) CPE have used remarkably coarse scales, e.g. 0, 1, 2, where others might be tempted to use 1, 2, 3, 4, 5 and an implied or even a finer scale with higher scores for increasing usefulness. The case for a coarse scale is that if you allow a finer scale then even experts disagree on the ranking and the alleged extra precision may turn out to be spurious.

The case against is the proven fact that final rankings tend to display a preference for the mean, arising because so many factors score 1 and cannot score 2, and in aggregate display little distinction between projects. In the future it is expected that the scales will expand, either by allowing a 5-point division on individual rankings (instead of 3) and/or by combining the final rankings to give a 0—30 scale instead of 0—14 as at present. It is clearly a matter for experiment.

Application to the Civilian Sector

There will be two possible uses in the civilian sector and four main difficulties.

The uses are

(a) Evaluation of technological proposals in a given field.

(b) Generation of technological work because the construction of a relevance network for a particular field shows that certain technical capabilities are desirable or necessary, if certain goals are to be met.

The difficulties are

(c) The degree to which economic systems, sub-systems, and technological capabilities interact will make it difficult to limit the size of networks even in a restricted field. It will, therefore, be necessary to be careful in taking decisions about the point where relevance stops and becomes really irrelevant even though a connection is logically perceived.

(d) The number of conceivable hierarchic levels will tend to

217

expand in certain areas, producing the necessity either to recompress the number of levels, or to expand the scoring system to rank the interconnections in a logical way. Honeywell have 8 levels to get down to technological deficiencies in their American PATTERN programme (see below) and de l'Estoile suggests 6 in his civilian application to get down to desired technological capabilities.

(e) The factors that contribute to attainment of goals in the economic sphere are not all technological. For example, export volume is affected by technological excellence but it is quite impossible to obtain sales without sales effort. Technology is necessary but not sufficient.

In the military context, technology is necessary and the remaining management requirements can almost be guaranteed provided the environment and the goals remain as predicted.

Given this fundamental difference, it will be necessary to take a view of the extent to which, if technology makes goals attainable, they will be achieved bearing in mind the optional nature of some of them.

(f) It will be more difficult to obtain a consensus about the goals toward which the civilian sector is aiming. There are many options and some depend very much on one's philosophy of life. However, the relevance network approach can be used to examine the technology requirement consequences of alternative directions in which various aspects of the economy might be aimed.

Having outlined the difficulties, it seems that they partly solve themselves if one supposes that, for evaluating actual proposals for research, it ought to be possible to use a contracted network, having fewer levels but just sufficient to estimate relevance and need to presumed goals. For predicting what future technological capabilities will be acquired and hence to show what research ought to be initiated, a more detailed relevance network can be drawn but it can refer to as narrow a field as possible. Thus, what it gains in extra levels will be restricted to those activities or capabilities which are relevant to the more restricted goals or missions.

As an example of the levels required for predicting needs, de l'Estoile quotes the following for transportation.

Goals. e.g. satisfy Government transportation policy.

Missions. e.g. guarantee at minimum total cost. Public transport to satisfy the market demand.

Functions. Here he lists a matrix of requirements for ranges of distance, e.g. 0—0.5 km, 0.5—5 km, 5—50 km etc. and for differ-

218

ent classes, e.g. business and pleasure travel, and goods in lots of less than 5 tons, 5—100 tons and more than 100 tons. This then is filled in as required by objectives.

Objectives. These are specific tasks fulfilling a function such as transporting 5000 people an hour at 250 km/h door to door over 100 km where the objective is based on known requirements, basing volume on geographic studies and desired speed on theoretical studies of type done by Batelle Institute, Geneva (see Fontella, 1969). The next levels are describing means to meet the objectives.

Sub-systems. These refer to specific aspects of the desired operation—propulsion, baggage handling, access etc.

Elementary technologies. or technological capabilities, are the link between basic research projects and the sub-systems and they specify the type of alternative proposed to satisfy the sub-system, e.g. propulsion by picked-up electricity, engine run by carried fuel, linear induction motor.

Basic projects. These study specific proposals about aspects of the technological capability.

The precise terms used for each of the 7 levels could be changed to suit a given subject, only experience will show how to do the logical breaking down.

De l'Estoile quotes criteria he suggests for the transport relevance network and these are:

(1) Mission dispensability
   0 = indispensable to
   4 = very little use
(2) Function assistance to mission
   0 = condition attainment of mission to
   2 = could possibly help
(3) Objectives
   (A) current realisation
      0 = does not exist to
      2 = almost established now
   (B) Dispensability to the function
      0 = conditions attainment to
      2 = could possibly help
   (C) Breadth of impact
      0 = affects many people to
      2 = interests a few.

These three criteria link up to form one index of importance of trying to achieve the objective.

(4) Sub-system necessity (based on systems analysis)

   0 = indispensable to

   2 = useful in some circumstances

plus an index of the likelihood that the sub-system (a given way of achieving something) will actually get used.

   0 = virtually certain to be adopted to

   2 = limited likelihood of being used

(5) Technological capability

The suitability of the particular capability to provide the required sub-system, ranked according to criteria which must be devised for the task which has to be done. Thus, in transport one might make an index combining scores against the attributes

(A) cost

(B) safety

(C) comfort

(D) convenience (frequency etc.)

(E) nuisance (dirt, noise etc.)

The principles of combination would be as shown in Annex A.

(6) Basic project effect on the technological capability

Here de l'Estoile uses some combination of the likely relevance of the project to achieving the capability required, without costing too much. This needs suiting to the technology concerned and the rank will probably be from

   0 = must be studied in any case to

   2 = will help in attaining the desired capability

All these rank numbers must be combined in an agreed way to give an overall ranking of utility towards the given goal or mission. This ranking will then combine with another which combines

(A) Cost considerations whether they are likely to be excessive or not.

(B) Probability estimates that the project can achieve the desired end within, before, or after the desired time of implementation.

The two rank numbers of utility and cost—risk combined give the overall indication of the desirability of initiating a given research task. Alternatively, if the research task is not contemplated, the relevance network ranking for each technological capability indicates the relative desirability of being able to accomplish that capability. This gives the indication that research, not otherwise proposed, should, in fact, be initiated to procure the capability.

Emphasis is made on the need for a multi-disciplinary approach at the objective stage. The narrow approach will merely suggest

220

better and better steam engines. A wide-ranging approach will foresee innovations like hovercraft as a principle during the steam-engine age.

ANNEX A

Combining Rankings with Respect to Different Attributes

The combination method used is necessary because it is implicit that ranking is not on a cardinal scale and that it is impossible to be explicit about the relative weightings of different attributes. Simply stated, it is necessary to decide the preference order for two projects ranked against attributes A and B given that their rankings were

Project I	good A	neutral B
Project II	neutral A	good B

Supposing these combinations were ranked equally, it is then necessary to decide whether the same equality exists between projects ranked thus

Project III	good A	bad B
Project IV	bad A	good B

It is clear that project I is preferable to III and II is preferable to IV but ranking III against IV may invoke special attitudes to relative combinations of goodness and badness. To illustrate, using numbers to denote the scale

$3 = good$
$2 = neutral$
$1 = bad$

project I is $(3, 2)$ and project II $(2, 3)$ and if these are equal they can be represented by $3 \times 2 = 6$ or $3 + 2 = 5$ for each project. For projects III and IV we cast doubt on whether $(3, 1)$ is the same as $(1, 3)$.

We can construct a network of lines connecting the possible combinations of scores as shown in Fig. A1. The arrows indicate preferences which must exist if the two attributes are weighted

221

$$3, \quad 3 \to 3, \quad 2 \to 3, \quad 1$$
$$\downarrow \qquad \downarrow \qquad \downarrow$$
Attribute A $\quad 2, \quad 3 \to 2, \quad 2 \to 2, \quad 1$
$$\downarrow \qquad \downarrow \qquad \downarrow$$
$$1, \quad 3 \to 1, \quad 2 \to 1, \quad 1$$

Figure A1

equally and uniformly. If A is more important than B, however, the preferences can be written as in Fig. A2.

$$3, \quad 3 \to 3, \quad 2 \to 3, \quad 1$$
$$\downarrow \qquad \downarrow$$
$$2, \quad 3 \to 2, \quad 2 \to 2, \quad 1$$
$$\downarrow \qquad \downarrow$$
$$1, \quad 3 \to 1, \quad 2 \to 1, \quad 1$$

Figure A2

It is now necessary to decide whether the relative weightings are such that there are any equivalences or whether there are intermediate shades of distinction which are worth recognizing. For example, in Fig. A2 there are two progressions which must be preserved, viz.

$$3, 2 > 3, 1 > 2, 2$$
$$3, 2 > 2, 3 > 2, 2$$

but is 3,1 to be distinguished form 2,3? Firstly is it possible to distinguish and secondly is it of some value to make a distinction?

In the CPE work it was assumed that there was no difference in weight associated with most of the factors, i.e. they were mostly 0, 1, 2, ∞ and furthermore that for pairs of rankings $(2, 0) = (0, 2) = (1, 1)$. This makes life very simple since combination of two factors merely means adding rank numbers which makes the matrix layout seem ponderous. However, there is nothing to prevent one having more subtle combination rules which do require a matrix. The only condition is that they be self consistent. Laying out all possible combinations as a matrix is necessary before one decides whether a simple combination rule like the addition used by

222

Index of economic relevance	Index of real utility															
	0	1	2	3	4	5	6	7	8	9	10	11	12	13	14	∞
0	0	0	1	2	3	4	4	5	5	6	6	7	7	7	8	∞
1	0	0	1	2	3	4	4	5	6	7	7	8	8	8	9	∞
2	0	0	1	3	4	5	5	6	7	8	8	9	9	10	11	∞
3	0	0	1	3	4	5	6	7	8	9	9	10	11	12	13	∞
4	0	1	2	4	5	6	7	8	9	10	11	12	14	14	14	∞
5	0	1	2	5	7	8	9	10	11	12	13	14	14	14	14	∞
6	0	1	2	6	9	10	11	12	13	14	14	14	14	14	14	∞

Figure A3

de l'Estoile can be adopted. If it cannot, the matrix must stay. If it can, then it is best dispensed with and people just add up four or five numbers.

As an illustration of non-additive combination rules, Fig. A3 shows the combination of military utility, u, plus cost risk, q, on a scale 0 to 14, combined with economic relevance, E, on a scale 0 to 6 assuming that, for very important military projects, economic factors were ignored.

In the CPE system, interconnections are limited in the interests of simplicity so that the scores which have to be combined are as shown.

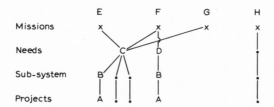

Any project is relevant to only one sub-system, and sub-systems are relevant to only one need. Multiple relevancies were overcome, it is assumed, by re-creating sub-systems, or ignoring multiple relevance. Needs were allowed to be relevant to more than one mission and an a priori rule for deciding a net index of need was worked out. It obviously requires testing to see how to adapt it to the civilian context.

Briefly the rule is as follows.

Let missions, E, F, G have rank numbers e, f, g. Let the relevance of C to each mission be C_e, C_f, C_g. The total relevance of C is made up of three rank numbers $(e + C_e)$, $(f + C_f)$, $(g + C_g)$ and it is proposed to represent their combined value by another rank number k which is given (completely empirically) by the rules

Let j = the minimum of the n rank numbers $(e + C_e)$, $(f + C_f)$, $(g + C_g)$. If

$$n.j + 1 \geqslant \sum_{i}^{n} (i + C_i) \tag{1}$$

i.e.

$$3j + 1 \geqslant (e + C_e) + (f + C_f) + (g + C_g)$$

then let

$$k = j - 1$$

if not let

$$k = j$$

What this means is that if one has, say, four rankings equal to 4, 5, 7, 7, then their overall worth[1] is 4 because

$$4 \times 4 + 1 \not\geqslant 4 + 5 + 7 + 7$$

but if one has four nearly equal rankings 4, 4, 4, 5 then their overall worth is equivalent to a rank of 3 because

$$4 \times 4 + 1 = 4 + 4 + 4 + 5$$

In practice, this rule means that for two relevancies ranking one integer apart, e.g. N and $N + 1$, the final rank is $N - 1$. This follows because, taking $j = N$,

$$2 \times N + 1 \geqslant N + (N + 1)$$

Therefore the final rank, $k = j - 1 = K - 1$.

This seems a bit too much promotion! To make a combination

---

[1] The reader should remind himself, here, that de l'Estoile has adopted a perverse scoring system with low scores indicating greater worth!

of 2 and 3 into 1. However, any other rules developed would have to be done on the basis of argument and the CPE rule can only be assumed to serve their purposes.

Final score for military utility is found by adding the modified index of need derived from the above rule to the index of relevance of sub-system to needs and of projects to sub-systems. The military index is added to the index of cost and risk and Fig. A3 shows how these combine with the index of economic relevance.

ANNEX B

Cost—Risk Factor and Economic Factor in the CPE Work

The risk element is primarily concerned with estimating whether the ultimate objective to which the research is aimed will be attainable within the desired horizon. The sort of penalties associated with different times are

1970	1.0
1975	0.8
1980	0.5
1985	0.2

It is recognized that time of maturity is partly a function of cost and in order to give an impression of the way in which funding might influence time of success, a chart as shown has been suggested.

Cost or expenditure	Time to success		
	2—5 years	5—10 years	>10 years
5— 10			
10—100			
100—500			
500 +			

In each square one is asked to write a probability using levels of the type

zero
small
one in two
high
certain

It has been suggested that the table be filled up considering military utility and then considering economic utility.

The actual method of combining the cost—risk information with the military and economic is under review and the above are really only proposals. In the de l'Estoile paper, the weighting given to the cost—risk factor, q, is low (i.e. 0, 1, 2 compared with 0 to 11 for the military utility index). There seems to be a need to adapt the weight of cost and risk to the importance of economic objectives, since clearly it could be more important consideration than in the military context.

Economic Factors

There is a total of 7 factors against which projects can be assessed, but no project is assessed against more than 4. The combination of factors is suited to the type of economic effect the project will have. The factors are as follows.

(A) percentage of potential market that the military demand represents,

(B) potential field of application in terms of potential impingement in some way on a proportion of the CPN,

(C) the importance research plays in determining progress in a field,

(D) progress made in R & D capabilities,

(E) contribution to the capital of knowledge,

(F) contribution to institutions, e.g. to the contractor who will do the work, and

(G) potential effect on costs of production.

For projects affecting products

E and F, plus C or D, plus E, G, or B.

For projects affecting processes

E and F, plus C or D, plus B

For projects where major contribution to the economy lies in its contribution to fundamental knowledge the factors are

E and F, plus D.

This means the last project has inherently less economic influence than the other two categories.

There has been considerable discussion about the economic factor in the CPE work. It is also of interest to note that balance of payments and national independence have been considered as factors but are not currently in the list.

The logic of using a purely additive law for the derivation of final relevance scores is highly suspect. Even if it is allowed that the scores 0, 1, 2...........∞ can be considered as logarithms of subjective factors, then a purely multiplicative law exists and no calculation of expectance of success, in a complex situation, would be either purely additive or multiplicative. I have not attempted to calculate the total number of subjectively quantiified estimates which have been amalgamated by CPE into the final score but if one imagines the computer at work processing the scores from different assessors who are probably not even in good communication with each other, then one has some temerity in hoping for a final meaningful result. Furthermore, innovatory proposals which are not likely to fit into the rigid structure of "levels" (goals to elementary projects) will receive a stultifying score (unless some genius "bends" the system by constructing another "hierarchic level" solely for the purpose of including the innovation, and "organization" men are not prone to such creativity). This last point is extremely serious because a new proposal could have embedded within itself a new goal.

The relevance system does not allow explicitly for some of the very significant qualitative factors which are so important in the conversion of a new proposal into an output which incrementally enhances a goal. We all know that there are such factors which usually enter the decision making process and are dealt with by the decision maker through the use of an important part of his intuitive repertoire ("gut feel"). These factors include relative timing, available finance within a portfolio of separate expenditures, and the quality of the project leaders and their teams.

Although de l'Estoile's method was thoroughly investigated at a time when the U.K. Programmes Analysis Unit thought there was fairly urgent need for some such aid, the PAU have not used the method or any derivation of the method.

For comparative purposes a relevance method developed in the United States, to fill another real need in complex situations, will be described next.

An American Relevance Technique

The following short account of Honeywell's PATTERN scheme

(Planning Assistance Through Technical Evaluation of Relevance Numbers) has been heavily culled from Jantsch (1967) and Sigford and Parvin (1965).

The object of Honeywell's Project Pattern is, according to Sigford and Parvin "to determine which of all the attractive areas of advanced development begging funding support are the most relevant to our nation's needs. Which technical problems, if solved would advance our national goals? This should indicate intrinsically the work that eventually will have to be done."

Jantsch uses the following iconic diagram to show the processes contributing to the computer diagram.

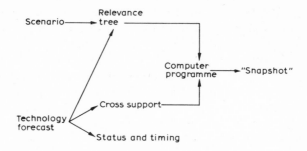

The qualitative scenario describes the broad objectives and goals of the United States in the particular field under review to a forward time "horizon" of about fifteen years. "It brings together the common background of knowledge and opinions, much of which is necessarily subjective, of the various members of the group and of the national experts consulted for this purpose. Upon this scenario were based the value judgements made on elements at the higher levels of the relevance tree." (Sigford and Parvin)

"At the same time, a technological forecast is made at the primary systems level and lower levels, aided by massive trend extrapolation and envelope curve techniques as well as other forms of qualitative and quantitative exploratory forecasting. Apart from an identification of primary systems, secondary systems, and functional sub-systems and their relationships used for the relevance tree, two sets of characteristics are assessed explicitly: cross support, which means spin-off to other areas, or general technological growth to be expected from tackling a specific technical system; status (research, exploratory development, advanced development, product design, availability) and timing for systems and sub-systems." (Jantsch)

"The relationship between technological deficiencies and the national objectives is shown by a carefully development relevance tree. The eight levels of the tree begin at the top with the particular national objectives of military and scientific pre-eminence (selected as targets for this project). The next lower level (and each level thereafter) contains those elements which, if our capability were upgraded would contribute to the element above. Hence, the second level includes three types of "conflict" (1) active hostilities, (2) non-combat operations, and (3) exploration of earth and space. The third level is "Forms of Conflict" containing the elements

Strategic global
Tactical theatre
Counter insurgency
Intelligence
Arms control
Area control
Exploration of earth
Exploration of space

The fourth level, "Missions", contains 46 elements such as

Destruction of enemy strategic resources
Protection of friendly resources
Population control
Exploration and use of hydrosphere
Exploration of planets

The eight levels and some sample elements at each level are shown in the following diagram of a relevance tree. The elements at the eighth level, then, list the nation's critical technological deficiencies as viewed by this group of people. It should be re-emphasized that many important non-critical factors such as lower cost and weight are not within the immediate scope of the project." (Sigford and Parvin)

"The criteria (α, β, γ etc), their weights (q_α, q_β, etc.) are estimated on the basis of the scenario. The setting up of these matrices is a major effort requiring a synthesis of expert judgement (Honeywell used 20 full-time experts during six months and their judgement was reinforced by contribution from experts in different departments)." (Jantsch)

Two normalizing conditions are introduced to assure the homogeneity of the logic

$$\sum_{\lambda=\alpha}^{\vartheta} q_\lambda = 1 \quad \text{and} \quad \sum_{j=a}^{n} S_j^\alpha = 1$$

229

Criteria	Weights of criteria	Items on level i			n
		a	b	- - - - - -	
α	q_α	S_a^α	S_b^α		S_n^α
β	q_β	S_a^β	S_b^β		S_n^β
γ	q_γ	S_a^γ	S_b^γ		S_n^γ
$\vdots$	$\vdots$	$\vdots$	$\vdots$		$\vdots$
ϑ	q_ϑ	S_a^ϑ	S_b^ϑ		S_n^ϑ
		r_i^a	r_i^b		r_i^n

S_a^γ = significance number (How significant is the contribution of item a to criteria γ?)

r_i^a = relevance number of item a on ith level.

N.B. These must be derived by assuming that subjective guesses are nearly the same thing as statistical probabilities of mutually exclusive events $\Sigma p = 1$.

"At each of the seven levels 2 to 8, a matrix is set up to match issues against criteria. (To be precise, from level 4 downwards, one matrix is set up for each "family" on the level e.g. one matrix for each family of forty-six missions that come under one of the eight forms of conflict .... logically there should be one matrix at each level.) This matrix has the general form: .... (Jantsch)." The relevance number is defined as

$$r_i^a = \sum_\alpha^\vartheta q_\alpha S_a^\alpha$$

which looks remarkably like the expectance from a set of events with pay off q.

"The definition of the relevance number is an arbitrary one, but seems to represent a natural choice and well reflects the original

230

Level	No. of elements	Typical elements
1. National objectives	(1)	Military and scientific pre-eminance
2. Types of conflict	(3)	Active hostilities Non-combat Exploration
3. Forms of conflict	(8)	Earth Space
4. Missions	(46)	Exosphere Lunar Solar system Universe
5. Postulated system concepts	(151)	Interplanetary probe Unmanned orbiter
6. Functional sub-systems	(425)	G & N Checkout and maintenance
7. Sub-system configurations	(340)	Active/passive F.M. ranging and direction
8. Technology deficiencies	(1981)	Optimum sensor selection for planetary horizon scanning Equipment reliability after long shutdown in space

problem of analysing the contributions of one item to criteria of different importance." (Jantsch)

"The total relevance figure of a particular issue on any level is then obtained by multiplying upwards to the top of the tree (or down from the top to the level of the issue in question)...... The simplest formula for a total relevance figure, R, for a particular functional sub-system (on level 4) would be

$$R_4 = r_2 r_3 r_4 = \prod_{i=1}^{4} r_i \qquad \text{(Jantsch)}$$

Intuitively, PATTERN has attractions because there appears to be some numerate logic underpining the method: relevance numbers have the same algebraic form as the expectance (of success) from a set of mutually exclusive events and total relevance has the same form as the expectance from a set of independent stages, each of which must be successful. On the other hand, no real-life situation as complex as the NASA programme [1] or even a programme for a modestly capital intensive industrial process can be expected to have such a straightforward analytic form. For example, alternative ways of achieving success can be simultaneously applied and contributions can be made outside the hierarchical structure of the relevant tree. Furthermore, subjective estimates are *not* the same as statistical probabilities and vary from person to person and when subjective estimates from many sources are amalgamated, with the aid of a digital computer, into a final number one wonders whether decision makers have not succumbed to a "consensus syndrome".

The requirement for ten man years of expert effort buttressed by outside consultancy simply to set up the matrices is indeed high but is this only a real measure of practical necessity in these days of high technology?

Conclusions about the French and American Relevance Techniques

Because relevance techniques have been commissioned and used by reputable organizations (note the patronage of PATTERN, p. 227 in Jantsch (1967) and the ministerial support for de l'Estoile's method) it can be inferred that relevance methods have been found to fill a real need in truly complex situations.

Analysts of repute fall into two categories when appraising the value of such relevance techniques. Some perceive such manifest dangers in the use of the extent methods that they will not, at any price, countenance their use. Others pragmatically believe that, despite the dangers, the burgeoning complexity of forecasting and decision making in modern systems of high-technology warrants the use of relevance techniques which have some inherent imper-

---

[1] To which we might infer that PATTERN has made a useful contribution.

fections. Whether the reader approves the underlying rationale and detail of either of the methods as described is up to him alone, a reiteration of an old statement of mine is the best council I can leave him with on this point.

"Let us not be carried away by the present popularity of some, apparently logical, numerate techniques of forecasting in which the analyst abdicates his responsibility to the computer after ascribing hastily chosen numbers to relevant facets of the problem."

It is a remarkable fact that available documentation of relevance techniques does not place more emphasis upon the selection and the weighting of relative importance of normative goals. Even if the reader chooses to agree with the philosophy and detail of the described techniques, he should still question the goal-setting procedure. Remember the statement made above in the description of de l'Estoile's method.

"The relative values of the three separate goals were not established. This amounted to ducking the issue and relative values started only at the mission level."

A prima facie requirement for any relevance system is the clear, unambiguous definition and weighting of normative goals, and the larger the system concerned the more carefully must this be done. In the United States, a determined attempt was made to handle this problem through the services of the National Goals Research Staff (NGRS), located in the White House (see *Towards Balanced Growth: Quantity with Quality*, Washington, D.C., U.S. Government Printing Office, 4th July 1970)[1].

The statement made by President Nixon when he announced the establishment of the NGRS on 13th July 1969 contained the following.

".... I have today ordered the establishment, within the White House, of a National Goals Research Staff. This will be a small, highly technical staff, made up of experts in the collection correlation and processing of data related to social needs, and in the projection of social trends."
"The functions of the National Goals Research Staff will include forecasting future developments, and assessing the longer-range consequences of present social trends, measuring the probable impact of alternative courses of action, including measuring the degree to which change in one area would be likely to affect another, estimating the actual range of social choice, that is, what alternative sets of goals might be attainable, in the

---

[1] For the history of the attempts by successive U.S. Presidents to state national goals see Huddle (1971).

light of the availability of resources and possible rates of progress, developing and monitoring social indicators that can reflect the present and future quality of American life, and the direction and rate of its change, summarizing, integrating and correlating the results of related research activities being carried on within the various Federal agencies, and by State and local governments and private organizations."

"The first assignment of this new research group will be to assemble data that can help illuminate the possible range of national goals for 1976, our 200th anniversary. It will prepare a public report to be delivered by July 4 of next year and annually thereafter, setting forth some of the key choices open to us and examining the consequences of those choices."

The one and only report of the NGRS made in 1970, and mentioned earlier, is something of a disappointment [1]. Only a verbal structure for national goals was presented and, forgetting the omissions, as far as it went it was difficult to disagree with. However, it contained no operable mechanism for the erection of national goals and the goals were not defined with the clarity needed to make a national relevance system viable. Perhaps no sensible person would have expected it to do so and would have dismissed as grossly impracticable any relevance system which aims to assess the relative worth of differently directed actions when they are aimed at goals, the relative worth of which has not been assessed.

Here a touch of cynicism might be allowed to intrude. The suggestion is that relevance systems have, and only will, be used to quantify the usefulness of projects to missions which have already been agreed as necessary to some qualitatively demarcated national goal. It may be that the selection and weighting of national goals will always be subjectively and qualitatively handled by the prevailing political process. Certainly I have never seen any work which explicitly weighs national goals in a quantitative sense [2]. Yet man is born an optimist and I do not personally dismiss the idea that a new Newton or Leibnitz will appear to invent an acceptable social calculus. What I am sure of is that such chimerical aid to goal setting will, if it becomes tangible, draw upon the assistance of mathematics and not rely solely upon the support of computerized arithmetic or verbal sociology, useful as these may be on some occasions.

---

[1] Since 1970 no other report has emerged from the NGRS and the staff have been absorbed elsewhere (see Huddle, 1971).

[2] The nearest I have ever been to this is when watching film actor Peter Sellers portraying the mad, fictitious, scientific adviser to the President, Dr. Strangelove.

Social scientists have not been tardy in realizing that decision making on societal goals is not fruitful without the aid of social "indicators", which can be constructed from social statistics (see, for example, Bauer, 1966). However, even after agreement is reached, and this is not yet, on the correct choice of social indicators, there still remains the problem of how to calculate their overall relevance. In the Economist, 25th December, 1971, on p. 15 there is an attempt to rank countries in descending order of social attractiveness which jocularly highlights the problem.

Combinations of social indicators representing population density, divorce, early marriage, population per doctor, deaths from road accidents, murders, infant mortality, ratio of cars to people, proportion of 17-year-olds at school, proportion of dwellings with baths, ratio of telephones to people, tax, and economic growth are made according to purely additive laws. Although the scores attached to the social indicators range from 0 to 8 for the attractive factors, and -8 to 0 for the unattractive ones and not ∞ to 0, the system mirrors the de l'Estoile method!

The rules for amalgamation of social indicators into an overall score need very careful consideration. To demonstrate this in rather a naive way, consider a situation where the only two indicators to be combined involve two of the indicators used in the entertaining Economist article: ratio of cars to people, and deaths from road accidents. Collisions between cars per unit time will be roughly proportional to the number of cars multiplied by the average speed multiplied by collision cross-section. The number of people risking death per car—car collision will be nearly proportional to the number of people per car. Thus one would expect the additive combination of the indicators to be

$$\frac{\text{cars}}{\text{people}} + (- \text{ deaths from road accidents per unit time}) =$$

$$\frac{\text{cars}}{\text{people}} + (-) \text{ an increasing function of (speed} \times \text{people} \times \\ \text{collision cross-section)}$$

and any crude attempt to maximize this score by changing the number of cars will lead to the lunatic conclusion that one should increase the number because the actual number of cars in use does not enter the second term! Alternatively, characteristics of cars can be determined as a function of population and the propensity to acquire a car, i.e. car speeds, collision cross-sections, and safety

E.Q. rating	Social characteristic
130 plus	Dedicated to good works and to the service of others to the point of self-effacement or even self-sacrifice.
120—130	Dedicated to socially useful works, absolute refusal to act anti-socially but ego not suppressed.
110—120	Socially unimpeachable behaviour, balanced attitude between ego and social environment. Capable of unselfish behaviour.
100—110	Responsible and reliable in the right environment, but prone to accept standards of the majority.
90—100	Good citizen in routine conditions but capable of mean, selfish acts. Occasional liar.
80—90	Social being under supervision but capable of occasional dishonesty (not returning excess change, shop-lifting, etc.), poor sense of ethical values, attracted to lower standards, fond of "kicks".
70—80	Inclined to envy, hatred, occasional cruelty, and criminal behaviour. Prone to fall foul of the law.
70 minus	Brutish, malicious, cruel, habitual criminal.

can be determined by a function that is a consequence of the law used for combining the social indicators. Furthermore, the reader should ponder the consequences of choosing an ill-advised combinatorial method and at the same time neglecting an important social indicator, or relevance number, when calculating the overall score.

Characteristics of social behaviour are a major determinant in any technology assessment and arbiters suffering from hypobulia (a propensity toward psychoneurotic haste in making decisions) are not given enough guidance, by their analysts, on these factors. Dennis Gabor, 1971 Nobel Laureate in physics, has for some time been concerned that we should escape the "triple danger of nuclear war, overpopulation, and existential nausea born of inner emptiness". In Gabor (1966) he has pointed out that there is some need for a scale of ethical behaviour (E.Q.) to complement the scale of intelligence quotient (I.Q.). Gabor's idea is that, whilst I.Q. might be the best indication of problem-solving ability, there is some need for an E.Q. which will give some indication of the tendency to ask, and solve, questions which are aimed "at the views of the individual regarding others, not himself". Taking the Stanford—Binet tests of I.Q. as his model, Gabor suggests that there is some virtue in trying to rank the population according to their ethical performance, thus allotting E.Q. to them corresponding to the rank in a Gaussian statistical distribution with median 100 and the (traditional but otherwise arbitary) dispersion of 16.14. The outcome of such a measure would be that social planners could construct a sampling function of I.Q. and E.Q. (measured separately) which, in all probability, would be a two-dimensional Gaussian distribution. The hope is that such a joint sampling distribution of I.Q. and E.Q. would give the planner, the analyst, and arbiter-of-decision a quantitative picture of the available human material in any environmental situation. Definition of the "ethical scale" is certainly beyond my ken but to trigger thought Gabor has left us with his suggestion (Table A-1).

Other Methods of Technological Forecasting

The methods described above are easily categorized into normative or extrapolatory techniques of technological forecasting. Other methods which comprise the subject are not so conveniently placed. Techniques such as morphological analysis, mathematical modelling, Delphi forecasting techniques, cross-impact analysis, systems analysis and dynamic modelling, and intuitive methods of scenario writing can all assist either the extrapolatory or the normative forecasting process. Any of these techniques can be used to work from the present to the future, or backwards from a desired goal to the present.

Since mathematical and dynamic modelling have such universal application and play a large part in any technology assessment exercise, it is not necessary to describe them specifically in this section. For completeness, however, a superficial explanation of the remaining techniques is made below.

Scenario Writing

Historically, scientific prognosis of the course of technology has a respectable record. But each prognostication has, in the long term, been dependent upon the calibre of the forecaster or the actual technological innovator's perception. It is extremely doubtful whether any systematic effort can be made to solve the problem "How do we foresee technological innovations?". Therefore, forecasters should not become twentieth-century alchemists by exclusively attempting to turn statistics of retrospective events into an allegorical telescope through which they hope to perceive the future. Attempts to use the psyche of known or potential innovators are likely to be more effective than regarding technological forecasting as an amorphous source of pseudo-scientific theorems which presents an open field for dilettante exercise. Sophistry must be savagely restricted at its onset.

A start to using the psyche or seer-like qualities of the forecaster may be usefully made by asking him to write an imaginative essay on the possible future. Such a literary effort is described in technological forecasting parlance as a scenario.

Scenarios are often embellished with statistical evidence or projections but basically they consist of verbal predictions. The current high priest of the scenario-writing cult is Herman Kahn and apart from work by the great literati like H.G. Wells and scientists of Leonardo da Vinci's perception, the work of Kahn *et al.* (1968) provides the most instructive examples of the method.

The Delphi Method

Methods of amalgamating scenarios or other individual types of forecast into a whole, representative of plural opinion, need to be carefully selected. It was thought by the creators of the Delphi technique, Dalkey and Helmer (1963), that the anonymity of the contributors to the overall forecast should be preserved throughout the process of combining the individual contributions. The reasons for this were, essentially, that the exchange of opinion in a personal manner would be inhibited and biased by the extraneous psychological inter-relationships between those proffering forecasts. Any reader who has submitted himself to the ardours of committee membership will appreciate this point of view because the individual's contribution, in committee, is dependent as much on his and others' idiosyncrasies of communication as it is upon the intrinsic worth of the contribution he is trying to make.

The Delphi technique purports to remove the psychological pressures of direct confrontation with an iterative questionnaire process. Respondents are individually polled for their opinion on future events and at the end of each round of the questionnaire the aggregate results, usually in statistical form, are made known

238

to each respondent. He then has the opportunity, in the next round of responses, to modify his previous predictions in the light of the previous aggregate response.

As far as one can tell from the published results of Delphi Questionnaires (usually in the graphical form of distributions of probability that a particular event will happen at a given time), the iterative process results in a convergence of opinion. The probability distributions are normally uni-model (about a mean time) and very rarely do they have multi-modal shapes which would indicate minority or splinter-group opinions. Reasons for this may well be that the statisticians who calculate the probability distributions tend to truncate the upper and lower quartiles of the distributions, i.e. chop-off minority opinions, or that by informing a respondent that he holds a minority opinion and then asking him to explain why, the organizers force the "mavericks" into conformity.

Naturally, any Delphi Questionnaire is only as good as the quality of the respondents and much has been written about their selection. More importantly, however, the method intrinsically lacks an adequate means of discriminating between good and poor responses and it has been criticized, with some justice, as a means of obtaining a "consensus of the mediocre". Nevertheless, the analyst who has no a priori acceptance as a conductor of analyst—respondent interviews which would allow him to cross-fertilize information from one respondent to another circumspectly, may be forced to adopt a questionnaire technique to obtain information. (See Parker (1970) for a useful study of the U.K. chemical industry.) Furthermore, an argument in favour of Delphi can be made when, for some reason, there is a need to include such a large number of respondents that time does not allow personal interviews with all the advantages to be gained from the nuances of communication observed. This last argument is not, of course, acceptable to those who believe that only a few people have sufficient know-how to make accurate predictions. In this case Delphi will only be tolerated as a means of predicting future collective behaviour.

There is unfortunately a tendency to conduct Delphi questionnaires when the analysts cannot think of anything better to do. The advantage of this to the analyst is that he is seen to be doing something active, if not fruitful.

239

Cross impact analysis is concerned with the problem of producing an internally consistent set of subjectively quantified "probabilities" that an inter-related set of events will occur at stated times in the future. The method depends critically on the premise that a mathematical relationship determines the enhancement, or otherwise, of the probability of a later event occurring *if* there has been an enhancement in the probability that an earlier related event will occur. Generally, the assumption is that the mathematical relationship is quadratic and the process then involves the use of random number generators and the iterative recycling (thousands of times!) of all the quantified, a priori, subjective judgements (the "probabilities").

Monte Carlo programmes of cross impact analysis have their appeal to some analysts and James F. Dalby's clear description of the process in Cetron and Ralph (1971) is well worth reading. Nevertheless, until a cross impact analysis is verified or shown to be accurate in some ex post study conducted by analysts working in ignorance of the real-life outcomes, I cannot recommend the technique.

Morphological Analysis

Dr. Fritz Zwicky, the well-known astronomer, is the inventor of morphological analysis as applied in technological forecasting, see Zwicky and Wilson (1967) and Jantsch (1967). The word morphology is from the Greek *morphie*, meaning form, and *logos*, science; its meaning is therefore the study of organic form.

Morphological analysis invokes a rather pedantic detailing of all factors relevant to the problem. It relies upon an excessively time-consuming process of examining all the possible solutions to the problem in the hope that serendipity will occur and an outstandingly new solution will be revealed. The method of analysis relies upon a formal expression of all the pertinent factors in symbolic form and the placing of these factors into morphological "matrices"[1]. The following brief account of an application of morpho-

---

[1] The word "matrix" ubiquitously appears in the literature of technological forecasting and has become a standard part of the nomenclature. It is usually a misnomer because matrix algebra is rarely applied to the original table or "matrix". In most cases "tabular techniques" would be a more accurate description.

$$\begin{vmatrix} I_d & I_a & & & & \\ C_r & C_i & C_o & C_a & & \\ T_r & T_i & T_o & T_a & & \\ G_c & G_n & G_i & G_o & G_a & G_s \end{vmatrix}$$
Information. Digital or analog.

Carrier/transducer. Radio, infra-red, optical or acoustic.

Free transmission. Radio, infra-red, optical or acoustic waves.

Guided transmission. Cable, waveguide, heat-guide, lightguide, soundguide, and satellite.

Fig. 6. A simplified communications morphological master-matrix.

Sub-matrix for one kind of satellite communication system = $[I:_a \ C:_r \ G:_s \ \check{A}\hat{E}]$

Sub-matrix for demersal fish in a tropical ocean =
$$[L_{EE}+L_{SE}; H_N; C_{BT}; W_M; P_{DF}; B_{CP}; D_{PM}; T_C; DK]$$
(master-matrix not shown) where

I_a	denotes analog communication
C_r	radio carrier
G_s	guided transmission — via satellite
AE	environmental region
$L_{EE}+L_{SE}$	denotes ship borne and externally based fish location equipment
H_N	naturally occurring fish shoals
C_{BT}	bottom trawls
W_M	mechanical preparation of fish
P_{DF}	fish preparation by deep freezing
B_{CP}	fish stored in consumer packs
D_{PM}	waste disposal by conversion to fish meal and pulverising
T_C	transport by catcher vessel
DK	region undersea

logical analysis, see Medford (1968), should make the underlying principles clear.

Zwicky's morphological method of forecasting was adopted by the Programmes Analysis Unit after it was realised that a very wide sectorial analysis of marine technology required a systematic method of gathering information. The original intention was that the PAU should

(*1*) use morphological matrices to derive objectively all the conceptual systems which comprise marine technology; then

(*2*) examine the derived systems (including existing, immediately feasible, and speculative systems) for common features. An example of this would be the communality of sonar devices used for fish detection and sub-bottom marine survey instrumentation; then

(*3*) attempt a technological forecast of the ubiquitous features to determine areas of possible economic benefit which would determine normative goals for marine technology.

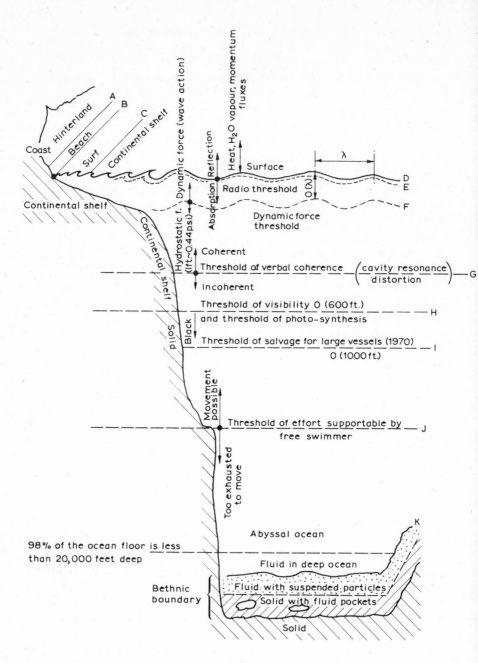

Fig. 7.

The first part of the exercise was an exacting but stimulating task. For each sector of marine technology a master matrix was constructed from which sub-matrices were derived, each representing a conceptually possible marine system within the sector (see Fig. 6). Each sub-matrix was then related to the environment by using a specially constructed system of denoting regions and thresholds of performance within the environment [1]. (See Fig. 7, which is self explanatory.)

The first part of the exercise yielded about twenty marine systems for transportation, twenty-two district systems of marine communication, and many thousand possible fishing systems. Naturally, the exact number of the sub-systems depends on the thoroughness with which the master matrix has been constructed because this number equals the multiple product of all the numbers on each row of the master matrix: $n_1 \times n_2 \times \ldots\ldots\ldots n_n$. An accusation levelled at morphological analysis is usually made for this reason, i.e. you only get the sub-systems that you put into the master matrix in the first place!

Even with the relatively limited number of sub-matrices in this example, the sheer weight of the problem of seeking common features handicapped the analysts to such a degree that it was impossible to achieve the second part of the exercise in the allocated time. Some retrieval of effort was, however, possible. It was discovered that in some sectors, particularly communications, the parameters forming the sub-matrices could be quantified to show what had been currently achieved and what was physically possible, thus providing a base from which scenario-type forecasts could be made; an example is given in Medford (1969). In other applications of the method, varying degrees of success may be possible (see, for example, Bridgewater, 1969).

Summary

The quotation from Thomas Carlyle which prefaced this chapter contained reference to Cagliostros and the ephemeral nature of their plausible forecasts. Cagliostros both great and small joined the technological forecasting boom. Some had their creative imagi-

---

[1] Such labelling of environments is not common and is possibly one advantage of the morphological method. Figure 7 is for illustrative purposes only and should not be taken as up-to-date.

nations harnessed and became useful members of assessment teams, others were detected in their quackery and moved to pastures new. There now remains a hard core of technological forecasters who survive because their efforts are badly needed in industry or the public service. Apart from consultants, however, it is significant that the remaining specialists are all attached to some other activity, such as market analysis, research planning and management, and technology assessment. In fact it is quite remarkable that the earlier forecasters did not generally realise that they could not maintain the credence of their subject without attaching it to industrial intelligence, and then attaching both to technology assessment.

Searching for unanticipated secondary consequences of technological and scientific innovations and considering feasible technological alternatives to those innovations which are thought to be harmful is now the meatiness of technological forecasting. It might even be fortuitous that forecasters have had so much time to practice their art before the call came for legislation of the technology assessment process. At least, now, the best efforts of the early technological forecasters give some substantial proof, and warning, of what is possible in trying to foresee and control the impact of technological innovation. Lord Cherwell might have been wrong in telling the House of Commons in 1945 that intercontinental ballistic missiles would not be competitive with bombers in the foreseeable future (see Ayres, 1969). But Von Karman was very successful, indeed, in predicting future trends in the development of all kinds of transportation (see Jantsch, 1967). Von Karman's scientific sifting of the speed/weight and power/weight ratios for all vehicles still indicate feasible limits of transportation performance whereas Lord Cherwell's prognostications could be mistaken for Carlyle's "forged notes".

One practical way in which the technological forecasters have prepared the ground for technology assessment is by forming associations where their subject can be discussed against the wider implications stemming from moves to legislate for technology assessment. I sincerely hope that organizations like the European Technological Forecasting Division of EVAF take the opportunity, presented by the current interest in Technology Assessment, to bring together industrialists, the technocrats in the public service, and members of the public to discuss future impacts of technology against a background of technological forecasts made for individual industries. Perhaps the spur of legislation for technology assessment will drive all compartmentalized technological forecasters together and away from the forecasting of isolated trends like industrial growth, employment, communications capacity, surgical trends etc., towards more generalized forecasts, thus avoiding the criticism of present deficiencies in forecasting voiced by Donald A. Schon in his mimeographed (undated) paper "The Problem: Forecasting and Technological Forecasting".

"We know enough about (total social—economic—technical systems) to understand that they are characterized by large numbers of interdependent variables; that these variables change according to differential rates; that small rate changes may be highly significant for the system as a whole; and that the system is characterized by dynamic feed-back, so that it is, in complex ways, self-controlling. (But) efforts at understanding technological change are now fragmented: economists tend to look at a smaller number of relevant factors, such as trends in output and demand relative to capacity; technologists and students of technology tend to look at technology as though it had a life of its own; sociologists tend to relate it to social systems characteristics. There is .... no current effort toward the development of a more nearly complete theory".

References

Ayres, R.U. (1969). *Technological Forecasting and Long-Range Planning*, New York: McGraw Hill Book Company.

Bauer, R.A. (1966). *Social Indicators*, Cambridge Mass: MIT Press.

Bridgewater, A.V. (1969). "Morphological Methods — Principles and Practice", in: Arnfield, R.V., ed., *Technological Forecasting*, Edinburgh University Press, p. 211.

Cetron, M.J. and Ralph, Christine, A. (1971). *Industrial Applications of Technological Forecasting*, London and New York: Wiley-Interscience.

Coleman, J.S. (1964). *Introduction to Mathematical Sociology*, London: The Free Press of Glencoe, Collier-MacMillan Ltd.

Dalkey, N. and Helmer, O. (1963). "An Experimental Application of the Delphi Method to the use of Experts", *Management Science*, 9.

Fisher, J.C. and Pry, R.H. (1971). "A Simple Substitution Model of Technological Change", in: Cetron, M.J. and Ralph, C.A., eds., *Industrial Applications of Technological Forecasting*, London and New York: Wiley-Interscience, p. 290.

Fontela, E. (1969). "Technological Forecasting and Corporate Strategy", in: Arnfield, R.V., ed., *Technological Forecasting*, Edinburgh University Press, p. 26.

Gabor, D. (1966). "Fighting Existential Nausea", in *Technology and Human Values*, Centre for the Study of Democratic Institutions.

Goodman, C.S. (1972). "Measuring Industrial Markets — Uses and Limitations of Available Data for Market Measurement", *Industrial Marketing Management*, 1, No. 3, 279.

245

Huddle, F.P. (1971). *The Evolution and Dynamics of National Goals in the United States*, printed for the use of the Committee of the Interior and Insular Affairs, 92 Congress, 1 Session, Washington: U.S. Govt. Printing Office.

Jantsch, E. (1967). *"Technological Forecasting in Perspective"*, Paris: O.E.C.D.

Kahn, H. and Wiener, A.J. (1968). *The Year 2000: A Framework for Speculation on the Next Thirty-Three Years*, New York: The MacMillan Company.

Lenz (1962) "Technological Forecasting", 2nd Ed., *Report ASD-TDR-62—414, Aeronautical Division Air Forces System Command*, U.S. Air Force, Wright Patterson Air Force Base, Ohio.

Lewandowski, R. (1972). "Long Range Forecasting Functions", *Industrial Marketing Management*, 1, No. 3, 378.

Medford, R.D. (1969). "The Application of Technological Forecasting" in: Arnfield, R.V., ed., *Technological Forecasting*, Edinburgh University Press, pp. 268—269.

Medford, R.D. (1970). *Further Remarks on Forecasting Techniques used in the PAU*, European Technological Forecasting Association, Paris, May, 1970, EVAF 39—40, St. James Place, London: or *PAU M. 15*, Programmes Analysis Unit, Chilton, Didcot.

Parker, E.F. (1970). *Chemistry and Industry*, January 31st, 138—145.

Pearly, R. (1924). *Studies in Human Biology*, Baltimore: Johns Hopkins Press.

Rose, Hilary and Rose, S. (1971). *Science and Society*, Harmondsworth, England: Penguin Books.

Sigford, J.V. and Parvin, R.H. (1965). "Project PATTERN: A Methodology for Determining Relevance in Complex Decision Making", *IEEE Transactions on Engineering Management*, March 1965.

Spencer, M.H. and Siegelman, L. (1964). *Managerial Economics*, Rev. Ed., Chap. 3. Homewood, Illinois; Richard D. Irwin.

Wood, Eric S. (1967). *Collins Guide to Field Archaeology in Great Britain*, London: Collins.

Zwicky, F. and Wilson, A.B. (1967). *New Methods of Thought and Procedure*, New York: Springer, p. 285.

CHAPTER 8

Technology Assessment of Curiosity-Motivated Research

"There are some who feel dismay that science, pure science, should have become so costly, that its cost should transcend what one individual nation, even the wealthiest, could hope to afford. I am afraid I do not share this dismay. On the contrary, if one considers the future of humanity, confined on a small globe, interdependent as never before, living off a limited biosphere, which until recently we have exploited with no thought either for our neighbours or for our children, if we consider that we also possess technological competence capable of providing plenty on a planetary scale if we were not hampered by the dead-weight of our preatomic and preglobal traditions, if you remember all this, one should rejoice in costly science, rather than feel dismay."

Prof. Abdus Salam (1971)
Nobel Laureate

Introduction

If technology assessment is ever to be Carpenter's effective tool, which will "facilitate appropriate societal control of physical inventions", its practitioners must be prepared to analyse some of the barely predictable outcomes of potential innovations stemming from pure, basic, or curiosity-orientated research (COR)[1].

---

[1] "COR" will be the designation used in this book for research which is not directed at a clearly defined economic or social goal. Readers interested in

247

At present, positive recommendations on constructive method-ology for the assessment of COR would obviously be premature because the multiple impacts of this kind of research have only recently come under extensive scrutiny. The purpose of this chapter is to fuel the debate on organization and control of COR and discuss the extent to which technology assessment techniques, "proven" on mission-orientated and applied research can be usefully applied to "purer" research activities.

Difficulties of Evaluating COR

Prognostication on COR compared with prognostication on applied research is as an intelligence exercise is to a market survey. Curiosity-motivated researchers are justifiably sensitive to outside surveillance. Attempts to foresee and force economic and social benefits from research discoveries, which could be embryonic innovations, are not given much encouragement by the researcher who hardly understands the ramifications of his own work. Much more effort must be spent by the analyst in familiarizing himself with the science behind COR than is customary in the assessment of applied research. A prime reason is that much of the science is new and the analyst, in the COR context, cannot disruptively appear as an assessor of potentially new technology, benefit—cost accountant, or a manager. Empathy in discussion of COR is best generated by demonstrating more than a shallow knowledge of the scientific principles involved.

Some aspects of technology assessment methodology as applied to mission-orientated research are obviously valuable in an assessment of COR, for example, communication and the cross-fertilization of opinion by a series of personal interviews and seminars, and the ability qualitatively to demarcate potential benefits in order to encourage, wherever possible, beneficial canalization of future COR. However, the detailed intracacies of formal benefit—cost analysis are not, it this stage, seen by most experts to be deeply relevant to an understanding of COR. Conversely, the intelligence aspect of assessment work (information collection and col-

---

(footnote continued from previous page)

semantic definition of the different catagories of research and the way these categories can feed innovation will find Langrish *et al.* (1972) a good source.

lation) is universally acknowledged to be *very* relevant, since COR has inherently much dependence upon synergy and serendipity.

Given that the assessor of technology is prepared to invest the necessary time to familiarize himself with the science researched and gain the confidence of those involved, in COR he has the imperative duty of performing a "look-out" function. When performing this role of "look-out" the analyst must, of necessity, perform the part of seer of future developments and this is where he is most vulnerable. To make a forecast of the future consequences of applied research is arduous enough, to make a technological forecast of COR is to openly risk ridicule. Nevertheless, some effort must be made to match the rate of growth of informed public opinion with the rate of growth of new knowledge arising from COR and, painful though it may be, the analyst has the duty to erect the best signposts he can, pointing to future benefits and disbenefits. Reaction to such prognoses is likely to come from the only experts available — the COR researchers with natural vested interest in maintaining, or increasing, funding and other support for their research. The sting in the tail of malevolent criticism is usually that the analyst would be better employed writing science fiction rather than attempting to assess COR prematurely. In the past, of course, journalists and authors of science fiction were indeed our only assessors of this "frontier" science and with the limited information available to them, they did a good job which was not always heeded. Now the time is ripe for something better and COR must be opened to more public communication and scrutiny.

At the time of writing, open publication on the problems of assessment of COR has tended to be excessively biased towards ex post examination of the impact old COR has had on currently successful, and unsuccessful, innovations in industry. Good examples of this kind of work are TRACES (1968), Gruber and Marquis (1969), SAPPHO (1972). Such studies have achieved great success in sparking debates on the value to society of COR and each has contributed empirical information on "historical" fact. On the other hand, structured analyses of the possible innovations yet to come from current COR are, in general, only contained in confidential commercial documents which make entrepreneurial bids for high-risk capital, documents supporting claims for research funds, and industrial and military documentation which examines "outre" ideas for signs of narrow sectorial benefit. All these analyses have, with reasonable excuse, a limited circulation and are not normally available for public dissemination and debate.

One way around the obvious restrictions on circulation of technology assessments of current COR would be to circulate information on the methodology used, or intended to be used, in assessment so that the general points of principle could at least be discussed. But the great uncertainty surrounding COR and the idiosyncracy of its parts means that, as with the ex post studies, the usefulness of such a generalized examination of current methodologies is limited. Notwithstanding the risks, a brave attempt by Byatt and Cohen (1969) was made in order to improve the methodology which could be applied to assessment of COR. Predictably, their proposed methodology would not even stand up against the evidence of ex post case studies although it did have a considerable catalytic effect on analysts who would assay one step further in the creation of method.

To this author's knowledge no generally acclaimed improvement has been made on Byatt and Cohen's rejected, heuristic, attempt at methodology. Therefore, in this chapter a cursory examination of the Byatt—Cohen Model will be made against one of the excellent case studies from TRACES (1968) and a speculative, but original, methodology will be proposed for calculating the notional benefits of COR. In addition, in order not to be hoist by his own petard, the author will masochistically reveal his own inadequate attempt at technology assessment of specific COR which could conceivably lead to the production of an X-ray (three-dimensional) holographic microscope.

Retrospective Study

An early worthwhile objective of Science Policy Studies, made under the auspices of the U.K Department of Education and Science, was to find means of identifying the economic benefits stemming from previously completed COR and use the findings to help make policy decisions on current COR. The Byatt and Cohen (1969) paper proposed a conceptual means of achieving this objective by means of a mathematical model of the innovation process. Their model depended upon constituent equations for the incremental change in discounted net economic benefit of a sience-based industry which could be brought about by notional delays or advancements in COR and/or information transfer.

In order to define constants in the Byatt—Cohen Model and test its applicability, a number of retrospective studies of science-based products and services was commissioned. Professor Jevon's Department of Liberal Studies in Science at the University of Manchester examined two of the Department's case studies on the role of science in innovation (see Langrish et al., 1972) and concluded "that the Byatt—Cohen innovations are in practise difficult to find, let alone investigate". Peter McLaren of the PAU and the author carried out contemporaneous studies and reached an almost identical conclusion.

Before considering general comments on the Medford—McLaren work done in an attempt to verify the Byatt—Cohen Model, it is useful to obtain some numerate understanding of what Byatt and Cohen were about. It is indeed salutary to examine one of the theoretical "pivots" of the original Byatt—Cohen Model in algebraic form, as follows[1].

$$\text{Change in discounted net benefit} = \Delta\{V_i \exp[-(n+T)r]\}$$

$$= \{\Delta(V_i) - rV_i\Delta(n) - rV_i\Delta(T)\}\exp[-(n+T)r]$$

where Δ's mean "change in" (thus $\Delta(n)$ is the change in n), V_i is the ultimate benefit of the science-based service or industry, r is the discount rate, and n and T the innovatory and information time lags. The longer representation of the right-hand side of the equation is not vital to the present discussion, it simply shows how the change in discounted net benefit is comprised of components dependent upon the changes in information and innovation times and the change in the magnitude of the undiscounted net benefit. For example, suppose the time lags in innovation and communication of information are reduced by two years each, i.e. $\Delta(n) = \Delta(T) = 2(-)$ (minus sign for a reduction), and the ultimate benefit is consequentially improved by 10% (through earlier access to information and marketing "know how"). Then we have

$$\frac{\text{Change in discounted net benefit}}{V_i \exp[-(n+T)r]} = \text{Fractional change}$$

$$= [0.1 + r(2+2)]$$

$$= 0.5$$

if, say, discount rate = 10%.

---

[1] This formula uses continuous discounting which is explained in Chapter 5 and its Appendix.

Because Byatt and Cohen originally used a discontinuous method of discounting and wrote the Δ's in partial differential form, their equivalent formula took on a more fearsome aspect which is illustrated below, not for artistic relief, but to demonstrate how a relatively simple concept can become more obscure to the non-specialist when written in special language.

$$\text{Change in discounted net benefit} = \frac{1}{d(n+T)}\left\{\frac{\partial V_i}{\partial(n+T)} - V_i\ln(d)\right\}\left\{\frac{\partial n}{\partial C_i} + \frac{\partial T}{\partial C_i}\right\}\partial C_i$$

where $d = (1 + \text{discount rate})$ and C_i is a cost of reducing time lag.

It will be noticed (more easily from the first formula and example) that the time lags are additive; thus it has been assumed that the process leading from COR is simply sequential (like the process illustrated in Fig. 2, of Chapter 5). That is to say that a sequence of separate steps COR, applied research, invention and development, prototype construction, market research, and sales leads to benefit. Extensions of formulas similar to the last above were recommended by Byatt and Cohen, the extensions to be made by multiplying the equations with "impressive" factors which purported to allow for statistical chances that the COR would eventually lead to industrial innovation, and relate true benefit to idealized commercial sales of the graphic form shown in Fig. 1.

Questions arising from the Byatt—Cohen formulation of their model are

(1) How is benefit apportioned, even in a simple chronologically sequential chain, when more than one piece of COR is involved?

(2) If the innovatory process is not simply sequential but multiple connected and sometimes parallelled in time, how does one

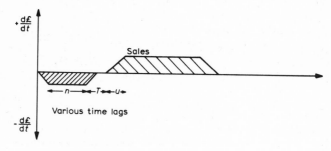

Fig. 1.

then calculate the likely contribution of the COR inputs and check on the sensitivity of discounted benefit to various time delays?

(3) Can simple statistical factors for the a priori probability of COR leading to industrial innovation be obtained for such a class of disparate events as that thrown up by ex post studies?

(4) For such a complicated thing as a market, can idealized commercial sales functions be accepted?

When the author, despite his misgiving at the excusable naivety [1] of the Byatt—Cohen Model, examined the ex post studies from TRACES (1968) for some evidence to corroborate Byatt and Cohen's assumptions, the most forceful impact was received from the multicolour diagrams which display with commendable clarity the chronological development of specific technological innovations. In TRACES, key events in COR, applied or mission-orientated research, development, and application are denoted as differently shaped nodules on a prolific dendritic growth map which depicts the historic cross-fertilization of apparently unrelated projects and disciplines.

Only in one of the five innovations studied in TRACES (Electron Microscope) does the dendritic growth ultimately converge to a sequence of steps from COR to profit. Starting from the data of the successful innovation and working backwards through time, it was found that the first mention of "basic" research occurs five years before the first commercial electron microscope was produced. This research was in direct line of development of the microscope and so does not appear to require consideration of parallel events at that time.

Looking at the definition of this event it is seen, however, that the "basic research", or COR, was carried out in order to investigate the phenomenon of emission. The important feature so far as it effects the innovation was the designing of an electron microscope to study this phenomenon in the laboratory. How much of the cost of this COR should be considered as cost to be offset against the later benefits gained from the electron microscopy?

Going further back in time, the position outlined in the TRACES case study becomes even more complicated. TRACES

---

[1] All first-order models of complicated processes tend to be naive. No personal criticism of Byatt and Cohen is intended by discussing their model in such a way. As may be inferred from an earlier statement, it takes courage to present a first-order model and the critics are often the beneficiaries.

considers the derivation of the Energy Distribution Law by Boltzman to be an important event in the chain of events leading to the innovation of the electron microscope. *But* this work has had important effects in almost every branch of physical science. Should all the benefits arising from successful innovations in Atomic Physics, Vacuum Physics, Electronics, etc. give some partial credit to Boltzman's work? What would have happened if Boltzman had not solved this problem? Would someone else have done so? Would the innovation of the electron microscope have taken place without Boltzman's Law? The simple answer is that we do not know.

Going back to the original problem of decision making on COR, "could we have known in 1870 that the work of Boltzman would have helped to make possible the development of a new instrument with an enormous range of use in metallurgy, physics, chemistry, biology, medicine, etc?". We think not.

An examination of any of the other COR events in the TRACES case studies reveals the same story. All that can be said is that each of the innovations examined has depended to a very great extent on a fund of knowledge accumulated as a result of curiosity-motivated research. Many of the COR projects have no easily identifiable connection with even the general field of the innovation. The well-known truism is that the value of the research result has occurred, very often, long after the COR discovery has been made[1]. This has required that there has been a successful, but labyrinthine, transfer of knowledge between the discoverer of the COR phenomena and the more practical inventor or innovator. It is not possible, in the general case, to take account of the cost of this transfer which may have been made across national boundaries, disciplines, and even decades of time: it is a tribute to human ingenuity that the transfer actually takes place. An additional factor which must be taken into account is that, in TRACES, only successful events have been considered. It is not

---

[1] Because COR information may have been obtained long ago in another place, it does not follow that society has not paid a collective price for it, or that society should not lay down good COR for future generations. An entrepreneur can certainly obtain benefit from COR without contributing fairly to its cost but this does not mean that COR is not important to innovation or that society should, say, reduce expenditure on COR. It can always be asserted that COR is likely to produce benefit, even in the extreme case when the outcome of research is the production of "malignant" knowledge. At least then one knows what one should avoid.

known how many people at what cost were trying to solve similar problems. Should the costs of unsuccessful attempts be included in a model? Ideally, some means ought to found for attributing "worthwhileness" to current COR but the method should not be exclusively comprised of the kind of accountancy which underpins the Byatt—Cohen Analysis and is normally used in benefit—cost calculations and penalizes, through discountancy techniques, benefits which occur in the medium-to-far future. Perhaps the labyrinthine process of useful information transfer takes a minimal time which cannot be reduced beyond a certain lower limit (although with present communication efficiency we are nowhere near such a limit now and great improvements can be made) and if we do not do the COR now perhaps we will not have the innovations we need in the "discounted" future.

No discussion of the pioneering methodology of Byatt and Cohen would be complete without some consideration of the following conceptual problems.

Is it fair to all other contributors to an innovation to attribute the net improvement in economic, or social, benefit to the originating COR? Surely all contributors, particularly in the simplified situation when one is lucky enough to find a simple sequence of clearly related steps, from COR through marketing to application, should be credited with some notional apportionment of the net benefit? Market forces prevailing at the time a step is completed determine, to a large degree, the return to the input factors involved. But in order to calculate the notional importance of COR, some notional apportionment of benefit to the different contributors must be attempted.

Assuming that the answer to the last question is "yes" and that the last sentence is accepted, how, then, do we apportion economic and social benefits to the contributing events?

In the author's opinion it is manifestly unfair to apportion the benefit in the ratio of the discounted cost of each stage of the innovation process. Some allowance must be made for the investment of intellectual and entrepreneurial skill. Contributors who invest capital on low a priori probability of success and then rely largely on intellectual and entrepreneurial effort should be correctly acknowledged.

In the next section a hypothetical method is developed for the notional apportionment of benefit which appears intuitively to be fair because magnitude of capital, entrepreneurial and intellectual, investment are all allowed for and later contributors are given credit for carrying on the success of their predecessors. Suggestions are also put at the end of the next section for the extension of this notional apportionment technique to parallel and multiple connected processes of innovation, similar to the realistic development profiles shown in the TRACES study. It is hoped that these tentative suggestions are encouraging to future methodology. They are not intended to stimulate the construction of large computer-based decision trees for the allocation of funds to COR. Very much more careful thought needs to be given to the effect of COR culturally, for future survival, and economically before methodology can be safely accepted into decision making in COR. For the present, a lead from the economists showing how COR, through "neutral"[1] and "non-neutral" technological change, affects the production of GNP (or some other superior indicator of communal well-being) would be very welcome.

Notional Apportionment of Benefit

As an indication of difficulties to be overcome in any project starting with COR and ending with a commercial profit or a worthwhile service or amenity, the quotient of expectance of total benefit to expectance of total cost is adequate. With a simple binary, success or failure, chronological sequence (like that illustrated in Chapter 5 on benefit—cost analysis) this quotient is, in the nomenclature of previous chapters,

$$\frac{\text{Expectance of total benefit}}{\text{Expectance of total cost}} = \frac{p_R p_D p_M \overline{B(r)}}{\overline{R(r)} + p_R \overline{D(r)} + p_R p_D \overline{M(r)}}$$

where the discounted costs of the various chronological processes are $\overline{R(r)}$ for research, $\overline{D(r)}$ for development, $\overline{M(r)}$ for marketing and manufacture, and $\overline{B(r)}$ is the discounted benefit, all calculated for a discount rate of r. The "probabilities" of success at the conclusion of each phase are taken to be p_R, p_D, and p_M.

---

[1] "Neutral technological changes include a change in the efficiency of a technology and/or a change in technologically determined economies of scale. A non-neutral change is associated with variations in capital intensity and the elasticity of substitution, of labour for capital." (See Muray Brown, *On the Theory and Measurement of Technology Change*, Cambridge University Press.)

If the assumption is that it is acceptable to apportion benefit and other credits to the contributing events of research, development, and marketing according to discounted costs of each stage, then notional shares of benefit would be apportioned in the following ratio.

$$\overline{R(r)} : \overline{D(r)} : \overline{M(r)}$$

But, as previously explained, such an apportionment would be grossly unfair to those who have contributed and risked capital and intellectual effort on low a priori probability of success.

How, then, do we allow for research finesse or intellectual effort when the results of innovatory research are so uncertain? An attempt to answer this question can only be made if subjective estimates of a priori difficulty are acceptable as probabilities (and these should ideally allow for serendipity and synergy). One method of notional apportionment would be according to the contribution made by each phase in reducing the appropriate uncertainty.

For example, the contribution of increments in the probability of success at each stage produces marginal enhancements in the benefit—cost quotient in the following ratio[1].

$$\overline{R(r)}\,\frac{\Delta p_{\mathrm{R}}}{p_{\mathrm{R}}} : (\overline{R(r)} + p_{\mathrm{R}}\overline{D(r)})\,\frac{\Delta p_{\mathrm{D}}}{p_{\mathrm{D}}} : (\overline{R(r)} + p_{\mathrm{R}}\overline{D(r)} + p_{\mathrm{R}}p_{\mathrm{D}}\overline{M(r)})\,\frac{\Delta p_{\mathrm{M}}}{p_{\mathrm{M}}}$$

At any time during the chronological sequence of probability, Δp can be obtained by subtracting the initial estimate of probability of success from the current probability. Thus on successful completion of the overall sequence, the values of Δp would equal 1—(initial value of p). (Commonsense needs to be applied if a notional apportionment of expectance of benefit is attempted before completion of the overall task *and* the a priori estimate of probability has turned out to be unduly optimistic because this would mean at some intermediate time that the value of Δp would be negative!)

The main concern here is the apportionment of achieved benefit (not expected benefit) so that the values of Δp are necessarily positive and the above ratios with $\Delta p = (1-p)$ give the notional

---

[1] This result is easily obtained by differential calculus; alternatively it can be derived by subtracting the quotient determined by probabilities (p) from the quotient determined by probabilities $(p + \Delta p)$.

shares of the achieved benefit. It is my opinion that such an apportionment is essentially fair because it allows for magnitude of both
capital and intellectual investment (the latter is likely to be proportional to $(1-p)$) and it also gives credit for carrying on the
success of previous contributors.

To consider what this means in quantitative terms, consider a
"typical" ratio of research to development to marketing and manufacture expenditures common to industry, i.e.

$$\overline{R(r)} : \overline{D(r)} : \overline{M(r)} = 1 : 10 : 100$$

and consider that a priori probabilities of success at each stage become larger as one moves forward chronologically through the processes

$$p_R : p_D : p_M = 0.5 : 0.8 : 0.9$$

Then, according to the hypothetical apportionment, the ratio of
notional shares of benefit would be

$$(0.5/0.5):(1+0.5 \times 10)(0.2/0.8):(1+0.5 \times 10+0.5 \times 0.8 \times 100)(0.1/0.9)$$

$$= 1 : 1.5 : 5.1$$

Which is quite a different result from the initial ratio of capital
investment!

In the limit of all a prior probabilities equal to unity (i.e. every
stage is certain of success), the ratio of notional apportionment of
benefit reduces to the ratio of the spends incurred up to the
completion of each stage, which is

$$\overline{R(r)} : (\overline{D(r)} + \overline{R(r)}) : (\overline{M(r)} + \overline{D(r)} + \overline{R(r)})$$

and this ratio bears some relationship to the actual pressure put
upon the management of each stage since an essential phase (linking what has gone before to the achievement of benefit) is under
pressure because of what has irredeemably been spent and what is
going to be the benefit after total success.

Return on Capital Investment in COR

Although, in real life, the suggested notional apportionment of

258

benefit has no real substance as payment, it is salutary to calculate the notional return on capital that one could fairly expect on COR investments. Before doing this it is important to realize that a programme containing a COR phase may commit capital which could be more safely spent in a conventional operation where a good of unconventional kind is manufactured or a conventional service is to be provided. It is instructive, therefore, to use the above notional shares of benefit and calculate the smallest acceptable (assuming "notional" shares become "real" shares) threshold benefit for the overall process.

Suppose that the smallest acceptable return on marketing and manufacture, set by the market is represented by the rate of return, m, and that the notional rate of return on COR capital is c. Then the simple situation of sequential contributions yields the following relationship between *sufficient* returns to COR and marketing and manufacturing capital[1].

$$\bar{R}(1+c) = \bar{M}(1+m) \frac{\bar{R} \Delta p_R}{[(\bar{R} + p_R \bar{D} + p_R p_D \bar{M}) \Delta p_M]} \frac{p_M}{p_R}$$

Substitution of the numerical values used earlier into the above relationship yields

$$(1+c) = (1+m)(900/46) \sim 20(1+m)$$

Thus we see that a moderate chance of success in COR of 0.5, given the acceptance of the notional share-out procedure, demands a return on COR capital greater than twenty times the lowest acceptable return on capital used for manufacture and marketing. If the same organization funds both the COR and marketing and manufacture, this is not a startling conclusion because, given a ratio of COR to manufacturing and marketing cost of 1 to 100, instead of recouping typical returns on, say, 100 units of expenditure, it must (in this case) recoup typical returns on 120 units of expenditure. However, the conclusion may be somewhat startling to companies solely performing contract R & D with pay-off

---

[1] These are returns *after* allowing for discounting, a comparison of internal rates of return is algebraically more difficult but the above mode lends itself to an approach which yields, the percentage return on the present value of investment.

linked to ultimate profits or to organizations buying R & D from without.

Naturally, with different ratios of $(\overline{R} : \overline{D} : \overline{M})$ and $(p_R : p_D : p_M)$, the ratio of the returns to the individual contributors obviously change. In general, $\overline{R} < \overline{D} < \overline{M}$ and the more curiosity-motivated the research the smaller p_R becomes. Thus the larger the ratio of returns (based on the notional apportionment of benefit) soars, the qualitative change is not surprising *but* the quantitative variation can be larger than is expected.

Threshold of Benefit for an Acceptable Venture involving COR

At some time during an assessment involving COR, before rigorous application of full analysis, it is prudent to examine the largest possible benefit of the overall venture. When the forseeable benefit is small, prudence obviously dictates an end to speculation unless the inferred social benefits balance the deficit. But how small is "small" in this context? It is possible to develop an empirical rule-of-thumb based on the foregoing analysis of a simple sequential process.

$$\overline{B(r)} \frac{S_M (p_R p_D p_M)}{S_R + S_D + S_M}$$

is the notional share of expected benefit accruing to the marketing and manufacturing process, where the symbols S are functions of the discounted expenditure and a priori probabilities of success (see above). Since it is reasonable to assume that the return to marketing and manufacturing must at least equal to the lowest acceptable return determined by market forces, we must have

$$\overline{B(r)} \geqslant \frac{(1+m)\,\overline{M(r)}}{p_R p_D p_M} \left(\frac{S_R + S_D}{S_M} + 1 \right)$$

Usually the ratio $(S_R + S_D)/S_M$ is less than unity and the following empirical rule emerges.

$$\overline{B(r)} \geqslant \frac{\overline{M(r)}\,(1+m)}{p_R p_D p_M} \times \text{(a number between 1 and 2)}$$

If the multiplying number is close to unity, the result is a familiar

260

one from ordinary commercial transactions but it is of some interest to note that one rarely needs to insist on more than twice the customary commercial benefit.

Where the probability of success in research is less than 0.5 the reader is warned that the above empirical rule of thumb should not be used. Furthermore, in *all cases* of excessively high expenditure on COR or development, or when a priori probabilities of either are very small, the penultimate formula should be used.

Further emphasis must also be placed on the fact that all the calculations of this section have been made on the assumption that management is firm in handling the project and that an unsuccessful phase will not be allowed to repeat itself. (See the examples in Chapter 5 on benefit—cost analysis comparing the different expectances of benefit and cost under good and bad management.) Management which allows repeated attempts of any phase without strict surveillance is so manifestly bad that my patience will not allow me to repeat the above analysis for the case when the a priori probabilities of success merely determine the expected time of completion of any phase. Students of escalation in mammoth technological ventures may, however, generate some insight by repeating the above analysis with forms of discounted expenditure which have integral representations with the limits determined by the inverse of the a priori probabilities (the mathematical process involves differentiating an integral whose limits contain the variable[1]).

COR which does not Sequentially Fit into a Development

Apart from assuming control by good managers, the above analysis is built up for a very simple sequential chronological process, viz.

COR → development → manufacture and marketing → benefit

Students of real innovations will, as mentioned earlier, realize that hardly ever has COR been successful in such a straightforward way. Research programmes are, for example, often duplicated or necessarily complemented by other COR programmes and a real

---

[1] $$\frac{\mathrm{d}}{\mathrm{d}p} \int_{a(p)}^{b(p)} F(r, t, p?) \, \mathrm{d}t = F \frac{\mathrm{d}b(p)}{\mathrm{d}p} - F \frac{\mathrm{d}a(p)}{\mathrm{d}p} + \int_{a(p)}^{b(p)} \frac{\mathrm{d}F}{\mathrm{d}p}(r, t, p?) \, \mathrm{d}t$$

chronological flow diagram is likely to have the appearance of a "spider's web" or the "Gordian knot". In theory such intricate chronological patterns are amenable to analysis and a few relatively simple patterns are analysed below:

(a) In the case of COR which is comprised of more than one necessary piece of research, each with its own expenditure and probability of success, the following simplex pattern could emerge.

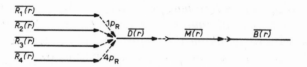

If all pieces of COR are essential to success in the research stage, then the overall probability of success in research is

$$P_{COR} = ({}_1p_R \; {}_2p_R \; {}_3p_R \; {}_4p_R) = \prod_{n=1}^{4} ({}_np_R)$$

and the associated research cost is

$$\overline{R(r)}_{COR} = \sum_{n=1}^{4} \overline{R_n(r)}$$

If the quantities P_{COR} and $\overline{R(r)}_{COR}$ defined above are substituted into the previous formula for notional shares of benefit instead of p_R and $\overline{R(r)}$, the shares between *total* COR, development, and marketing and manufacture are found.

Surprisingly, the individual notional shares of the total COR apportionment to the individual COR pieces of research contain no terms which allow for expenditure and are as follows.

$$\frac{\Delta_1 p_R}{{}_1p_R} : \frac{\Delta_2 p_R}{{}_2p_R} : \frac{\Delta_3 p_R}{{}_3p_R} : \frac{\Delta_4 p_R}{{}_4p_R}$$

Note that in sharing the total apportionment to COR between individual contributors, there is no allowance for the individual COR research costs! This does not mean that the hypothetical method of apportionment of benefit has broken down: it simply

262

means that, although one piece of COR may be very much more expensive than another, one cannot claim consideration for this spend if the expensive research is useless without the contribution from the cheaper piece of research. In the case where the individual pieces of COR are not under the auspices of the same organization and such an apportionment is to be attempted[1], the organization attempting the more expensive effort would be well advised to try to duplicate the cheaper, but necessary, piece of research. However, in setting up the example I have assumed that each contributor to COR can only handle his own special piece of research which is very often the case and usually means that finesse to achieve results in the cheaper, but unfamiliar, research takes such a time to acquire that imitation or duplication is out of the question.

(b) By way of comparison, consider the case in which COR consists of two independent pieces of research with success in either sufficient. Then the following chronological pattern would emerge.

If either piece of research is adequate, the success can be obtained with successful completion of either piece of COR or simultaneous success in both pieces. Thus the overall probability of success is

$$P_{COR} = {}_1p_R + {}_2p_R + {}_1p_R\,{}_2p_R$$

and total expenditure committed to both pieces at the onset is

$$\overline{R(r)}_{COR} = \overline{R_1(r)} + \overline{R_2(r)}$$

Substitution of these quantities into the first formula for notional shares of benefit will once again yield the shares between total COR, development, marketing and manufacture. The notional shares of the total apportionment to COR, made to the individual

[1] Vendors of essential research may not think in terms of a mathematical model but they certainly do look at their overall worth to the total COR effort and sell their contributions high if they see they are essential to collective success.

pieces of COR are given by

$$\Delta_1 p_R(1 + {}_2p_R) : \Delta_2 p_R(1 + {}_1p_R)$$

In this case at the completion of the total effort (through to benefit) the values of both Δp's need not necessarily be equal to $(1-p)$. If only one piece of COR is totally successful, the other piece can have a share attributed to it (the logic is that it has provided insurance against COR failure) but only if there has been some enhancement over the a priori probability. Naturally, if the value of Δp for the less successful case is zero, it has provided no insurance against failure, and its allocated share of benefit is zero. This last consideration does not arise in the last case because if any of the Δp's is zero in (a), when all contributary COR is essential, there is no benefit to share out.

(c) The final simple case is purely sequential but in this case the COR consists of two successive pieces of research with the second dependent upon the success of the first.

$$\overline{R_1(r)} \qquad \overline{R_2(r)} \qquad \overline{D(r)} \qquad \overline{M(r)} \qquad \overline{B(r)}$$

$$\text{------} \cdots \text{------} \cdots \text{------} \cdots \text{------} \cdots \text{------}$$

$$\qquad {}_1p_R \qquad\qquad {}_2p_R \qquad\quad p_D \qquad\quad p_M$$

and the quotient of benefit—cost is

$$\frac{\text{Expectance of total benefit}}{\text{Expectance of total cost}} =$$

$$\frac{{}_1p_R\,{}_2p_R\,p_D\,p_M\,\overline{B(r)}}{\overline{R_1(r)} + {}_1p_R\,\overline{R(r)} + {}_1p_R\,{}_2p_R\,\overline{D(r)} + {}_1p_R\,{}_2p_R\,p_D\,\overline{M(r)}}$$

which according to the original hypothesis yields notional shares of benefit in the following ratio.

$$\{\overline{R_1(r)}\}\frac{\Delta_1 p_R}{{}_1p_R} :$$

$$\{R_1(r) + {}_1p_R\,\overline{R_2(r)}\}\,\frac{\Delta_2 p_R}{{}_2p_R} :$$

$$\{\overline{R_1(r)} + {}_1p_R\,\overline{R_2(r)} + {}_1p_R\,{}_2p_R\,\overline{D(r)}\}\,\frac{\Delta p_D}{p_D}:$$

$$\{\overline{R_1(r)} + {}_1p_R\,\overline{R_2(r)} + {}_1p_R\,{}_2p_R\,\overline{D(r)} + {}_1p_R\,{}_2p_R\,p_D\,\overline{M(r)}\}\,\frac{\Delta p_M}{p_M}$$

In the above apportionment, the relative shares to the separate pieces of COR are made according to a ratio which is dependent upon the relative expenditures. The reason for this is (as opposed to the two previous cases) that the second piece of COR has the responsibility for carrying on the success of the first. (In the previous cases it was assumed that the separate pieces of COR were running concurrently in time and the expenditure on each piece was approved at the start of the whole process[1].)

Prospective Study

Prognostication on the future value of COR is, as previously indicated, a very risky business. In what follows, an account of a speculative prospective study of some current COR concerned with X-ray physics is revealed. The study was carried out by the author in the interstices of time between more urgently required and less speculative work done for the Programmes Analysis Unit. Therefore, responsibility for any inaccuracies or wild auguries is my own and no inferences should be made about the business of the old Ministry of Technology.

Choice of a sector of COR for the prospective study fell somewhere within the following matrix.

---

[1] Because of the way the quotient deals with the expectance of cost (in the denominator) it is explicit from the algebra that the management is efficient and will not commit expenditure before successful completion of previous stages. Should any stage fail and management call a temporary halt, there will be a new appraisal of the situation followed by new decisions which will be represented by a new benefit—cost quotient.

	1. Breakthrough imminent	2. Foreseeable breakthrough	3. No predictable breakthrough
A. Large sectors not directly sponsored by Mintech	A_1	A_2	A_3
B. Large sectors directly sponsored by Mintech	B_1	B_2	B_3
C. Small sectors not directly sponsored by Mintech	C_1	C_2	C_3
D. Small sectors directly sponsored by Mintech	D_1	D_2	D_3

In selecting the sector for study, the following criteria were used.

Exclude 1st column. Since any prognostication, after ascertaining the facts, is simply assessment of likely application which is part of technology assessment but not that part which is concerned with assessment of curiosity-motivated research.

Exclude 3rd column. Presumably the aim is not to end the study without even a chimerical prognosis about useful application?

Exclude 1st and 2nd rows. One man is not really capable of examining, part-time, large sectors of research (say the frontiers of low-temperature physics). It is highly probable that, in practise, he will not have the effort to collect even the ex post information.

Exclude 3rd row. Unless it is known that there is a small sector which one really believes that Mintech should support.

This leaves D_2, a small sector of research sponsored to some degree by Mintech, which has at least a partially foreseeable breakthrough.

Examination of work underway in Mintech circa 1969 revealed several published projects which could be grouped to form a very small research sector similar to D_2:

X-ray crystallography, spectrometry and microscopy

Typical work titles and quotations from the openly published Mintech literature were

Optical Metrology Division. "In an extension of optical instrumentation to the soft X-ray spectral region, a spectrometer is being made and will be used to assess the performance of diffraction gratings for this region...."

Mach—Zehnder Interferometry for the Soft X-Ray Spectral Region. An interferometer enables the "optics" of interference between wavefronts in the X-ray spectrum to be studied.

Inorganic and Metallic Structure Division. "To provide a service in the application of X-ray crystallography to the study of inorganic and metallic materials ...."

Some of the research into methods of using X-rays prevailing in 1969 were at such an unpredictable stage that it seemed fair to categorize them as COR or as on the borderline between COR and applied research.

Foreseeable Breakthrough for the Research Sector

Some of the information that gave birth to the thought that there might be a foreseeable "breakthrough" (or possibility of useful application arising from the research) is contained in the following random [1] selection of quotations.

---

[1] The use of the word "random" here is intentional. The need for a demonstration of how one would proceed with a prospective evaluation of COR arose so unexpectedly that there was no time to peruse systematically organized collections of information. The quotations given had been imbibed naturally through ordinary background reading (serendipity from the written page).

"The determination of the phase of a scattered wave has long been a significant and formidable problem for the X-ray crystallographer."

"A general applicability to X-ray microscopy of either the Kendrew or of the Buerger and Bragg method appears to have been restricted, in their present form, by the difficulty of ascertaining the phases of the various diffraction spots in the reciprocal lattice, in the general case."

"Stroke and Restrick noted the possibility of "lensless" Fourier-transform holography with noncoherent light to image-forming X-ray microscopy, especially because of the possible use of grating-like and beam-splitting properties of crystals."

"There are, however, several aspects in which an X-ray hologram microscope, if developed, would fill a role which cannot at the moment be filled by electron microscopes. Considerably greater penetrations without heating of the samples can be obtained with X-rays than with even very high-energy electron beams. This would be of a particular interest in areas such as metallurgy, and especially biophysics, in particular perhaps with live tissues and so on. It is also of interest to note that an X-ray microscope would not necessarily require a vacuum, which is a necessity with an electron microscope. Considerably greater resolutions might be obtained if holography were to be extended to gamma rays."

Professor George W. Stroke (1966)

It has been said, with reference to crystal structure analysis, that

"The researches of the past thirty years have built up a monumental edifice *but*. It is this *but* which mars the elegance of the whole subject, which can be summarized in the words phase angle and which forms the single barrier to the extension of single crystal Fourier methods to the complex domain of virus and protein structures, the chemistry of life itself."

A.D. Booth (1948)

"It is also possible that holographic methods could be used in structure determination. The suggestion is that a hologram be taken with X-rays and this be viewed with laser light so that it should be possible to "see" an image of the structure directly. Success in this project would be almost without parallel but the difficulties in producing spatially coherent X-rays seem insurmountable."

H.G. Jerrard (1969)

"We, International Business Machines Corporation, a Corporation organized ...., do hereby declare the invention for which we pray that a patent may be granted to us, and the method by which it is to be performed, to be particularly described in and by the following statements:
This invention relates to X-ray apparatus for examining objects by means of radiation in the X-band.
Present X-ray microscopes have a resolving power of about 2,000 Å ; that is .... The invention to be described hereinafter extends to the region of 100 to 200 Å, a ten to twentyfold increase in resolving power."

IBM (1969)

Serendipity first entered this prospective study when the above-mentioned IBM patent was discovered accidentally whilst browsing through the book reviews in a copy of the New Scientist. The people quoted above, and no doubt others of whom I am not aware, seemed to predicate the need for innovation. Moreover, Stroke and IBM appeared to have visualized a conceptual means whereby a possible breakthrough could be made which would lead to the needed innovation. It was, therefore, decided to make a study of previous COR, which has already led to present applications in X-ray crystallography and microscopy, and then look for potential future applications of current COR in the same field. Specifically, work began on assessing whether holographic (see the next section of this chapter for a description) and phase-contrast microscopes, with a resolution better than 200 Å, were a future possibility and could conceivably be of use in metallurgy, biochemistry, and medicine.

Initially, it was anticipated that the programme of study would proceed as follows.

Compile a cogent history of X-ray crystallography and microscopy, paying particular reference to work carried out in the U.K.

Predict the feasibility of a three-dimensional (or phase-contrast) X-ray microscope and the likely benefits such an instrument would have in fields such as biochemistry and medicine.

Recommend organization and financial structure of COR which would (hopefully) facilitate the approach to the goal of producing an holographic X-ray microscope and yield commercial and/or social benefits.

Because of the exigencies of other work (considered to be more important) the effort applied to the above task was mainly concentrated on the first two items, although it eventually became possible to outline a subjective proposal on future organization of COR in this field of endeavour. The following sections reveal, as far as possible, how the study in fact proceeded.

An Assessor's Quick Evaluation of the History of X-ray Crystallography and Microscopy

At the onset it was decided, not unnaturally, to avoid becoming involved in the historical ground work necessary to write a full chronicle. It was estimated that the amount of effort required to obtain the complete historical data would be enormous and would require an historian's expertise in the history of science (which I certainly have not got). A subjective annotation of events thought to be significant to the study were, therefore, made and these were buttressed with discursive writing on selected topics (mainly culled from Cosslett and Nixon, 1960) and Hosemann and Bagchi, 1962).

CARDINAL EVENTS

1895	Rontgen discovers emission of X-ray radiation at the University of Wurzburg.

1896	Burch	⎫		⎧ Botanical
1896	Ranwez	⎬ enlarged contact X-radiographs ⎨	Botanical
1898	Heycock and Neville	⎭		⎩ Metallurgical

1912	Friedrich, Knipping and Von Laue discover that X-rays can be diffracted by a crystal grating.
1914	W.L. Bragg formulates law of X-ray diffraction.
1915	Debye and Ehrenfest develop a method for analysis of atoms and molecules in the gaseous state.
1920	Bragg lays foundation of crystal chemistry by analysis of interatomic distances at University of London.
1922	Compton discovers corpuscular nature of X-rays, i.e. Compton Effect, at University of Washington.
1925	Duane and Havighurst determine electron density distribution.
1930	Pauling discovers homometric structures.
1931	Debye and Menke lay foundation for the analysis of substances in the liquid state.
1936	Silvert points out that the advantages of "point projection" for obtaining X-ray radiographs of high resolution.
1939	Von Ardenne suggests the use of electron lens for getting very fine focal spot.
1940	Ewald recognizes the importance of mathematical convolution operations to diffraction theory.
1948	Kirkpatrick and Baez construct an X-ray reflection microscope employing mirrors.
1949	Gabor develops new reconstruction principle in which information recorded with one type of radiation can be extracted by use of another type. Holography.
1950	Bormann discovers anomalous transmission of X-rays through reasonably perfect single crystals set for Laue diffraction, i.e. the Bormann Effect.
1951	Cosslett and Nixon build practicable X-ray shadow projection microscope with "point projection".

1951	Ramashandran and Thathachari build X-ray microscope based on Bragg Diffraction.
1952	Baez examines the possibility of constructing a zone plate for X-rays.
1952	Baez suggests that Gabor's method gives a better ultimate resolving power than that obtained with an X-ray microscope by direct imaging.
1956, 1958	Perutz and Kendrew use isomorphous replacement to overcome the "phase problem".
1965	Bonse and Hart report a successful X-ray interferometer which utilizes the Bormann Effect.
1965	North American Philips Company Patent Application on X-ray holograms.
1966	IBM Patent Application on X-ray Holographic Microscopes.

The limit of resolution of any microscope cannot be influenced by increasing the magnification. But Röntgen knew that resolution could be improved by decreasing the wavelength of illumination. Therefore there was an incentive to Röntgen, and others, to seek a usable refractive effect. Unfortunately, the refractive index (for X-rays) in likely lens material was less than one part in a thousand different from unity and construction of an early X-ray microscope was impracticable. Nonetheless, Burch and Ranwez and Haycock and Neville had utilized Röntgen's COR discovery, within a short time (see Cosslett and Nixon, 1960) by photographically enlarging contact X-radiographs of botanical and metallurgical specimens.

It may be significant, to those who believe that capability of the worker is at least as important as the supporting science and technology, to note that Otto Walkhoff's X-radiographic studies of fossil hominid material (made in 1902, see Brothwell, 1969) set standards of excellence which were not maintained during the following fifty years. Furthermore, metallurgical X-radiography, pioneered by Heycock and Neville, was allowed to lapse for about thirty years.

In the early part of the century it was obvious that ruled gratings for the diffraction of X-rays could only be made by extreme expedients outside the scope of current technology. The extremely short wavelength of X-rays, three orders smaller than the wavelength of visible radiation, apparently deterred experimentalists from successfully utilizing optical diffraction grating techniques.

Von Laue, however, had the sagacity to recognize that atoms of a crystal could act as "secondary radiators" of incident radiation and he suggested the use of crystals as diffraction gratings for X-rays.

Friedrich and Knipping provided experimental verification of Von Laue's ideas and then the Braggs showed how the intensity of diffraction patterns produced by crystals could be used to determine atom position in lattice structures. Subsequently, Duane and Havighurst showed how the electron distribution of atoms could be determined.

Succeeding theories of X-ray diffraction of liquids and of amorphous matter and gases were developed by Zernike *et al.* and Debye respectively and thus a range of methods for the X-ray analysis of the "micro" structure of matter was completed.

Since Röntgen's discovery and the above-mentioned pioneering "follow through", a very wide field of X-ray structure analysis has blossomed with practical advantage to industry through application in metallurgy, chemistry, biochemistry, physics, mineralogy etc. Nevertheless, these formidable gains have not been as easily won as might be imagined because nearly all applications are based on Fourier-transform interpretations of the intensity distribution of diffraction patterns which do not uniquely determine the diffracting structure. Indeed, Pauling discovered that there can be many interpretations of a diffraction pattern, called "homometric structures" and that only one of these interpretations will be the right one.

One straightforward way of unequivocally determining a unique structure would be to determine the *phase* as well as the *intensity* of the diffracting pattern. But phase is not easily determined and researchers have been forced to use great ingenuity to solve the "phase problem". Perutz and Kendrew showed that in proteins it is possible to attach heavy atoms (e.g. mercury, silver, lead etc.) in each lattice cell, without disturbing the conformation of the original crystal. Because the isomorphously replaced atoms of the higher polymers are of small atomic weight, their heavier replacements produce clearly separated peaks in the convoluted square [1] of the diffraction pattern and it is then possible to calculate the phase of the undisturbed structure directly.

Dorothy Hodkin, the Oxford Nobel Laureate, used the method of isomorphous replacement in her thirty-four year investigation

---

[1] A tedious mathematical operation that can be performed with analogue or digital computers.

of the structure of insulin but was handicapped because "there are no centric terms where changes in the X-ray diffraction pattern would be simply related to the positions of the heavy atoms". After achieving success and discounting early failure to introduce heavy atoms, Hodkin's co-worker Dodson (see New Scientist, 1969) made the prophetic statement: "with the increased experience we now have, and better equipment, the solution of protein structures can be reasonably considered *in terms of years rather than decades*". So we see that, even allowing for the introduction of digital computer backing and the technique of isomorphous replacement, determination of crystal structure is still a painstaking business[1] beset with "phase problems" and anomalous dispersion effects.

Whilst great strides, relative to the magnitude of the problem, were being made in the analysis of diffraction by matter, other researchers were tackling the experimental problems of building X-ray microscopes. Engström was extending the contact method of X-radiography by using finer-grained emulsions and larger photographic enlargments of the negative and had achieved resolutions of 1 μm (10,000 Å).

Point projection microscopy, using a finely focussed electron beam as a source of X-rays (after Von Ardenne) was being developed at Cambridge by Cosslett and Nixon and despite limitations of exposure time, contrast, and Fresnel diffraction, resolving powers of 0.5 μm were made possible. Compound mirror microscopes, using reflection of X-rays at glancing incidence, were also being developed by Kirkpatrick and Pattee and resolution exceeding 1.0 μm was obtained.

Of greater significance to this prospective study of COR, however, Gabor (1949) had fathered "holography" in an attempt to overcome early limitations of electron microscopes. In a nutshell, Gabor invented a means of "freezing" a three-dimensional wavefront of radiation onto a photographic emulsion and then reconstructing the wavefront, with its three-dimensional properties, at leisure. In essence, Gabor used a reference beam which interacted with the wavefront and, given sufficient coherence, produced an interference pattern on the emulsion[2]. The recorded interference

---

[1] Anyone who doubts this should read James Watson's literary masterpiece *The Double Helix*.
[2] Called a hologram — "entire recording". For his work on this invention Professor Dennis Gabor was awarded the Nobel Prize in Physics in 1971.

pattern when illuminated with the reference beam (or another "coherent" beam of similar spatial characteristics but different wavelength) produced a facsimile of the original wavefront with magnification, in the ratio of wavelengths, if the reconstructing beam was of greater wavelength than the original reference beam. Gabor's pioneering work was prompted by applied research directed at the task of correcting spherical aberration in electron lenses, in order to overcome the 5 Å limit in the resolving power of early electron microscopes (1948). It should not be thought, however, that Gabor's research was purely applied since, from boyhood, Gabor had nutured the idea of "lenseless cameras" and his research was directly in line with COR carried out by Abbe (1873), Topler (1867), and the Nobel Laureate (1953) Zernike's work on phase-contrast optical microscopes (1934). Unfortunately, Gabor's holographic techniques were not, at the time of discovery, directly of use in the development of electron microscopy because improvements in diffraction and reconstruction resolution could not then be made (the Laser was only invented later by Maiman and Schawlow) and, in the event, aberrations were ultimately overcome by better lens design. We should, nevertheless, note that Gabor's "electron microholography by two beam method" is certainly not forgotten or neglected: Möllenstedt and Wahl (1968) (Tubingen) and Tomita *et al.* (1970) (Kofu) have continued Gabor's work and Tomita shewed reconstructed images of magnesium oxide crystals (600 Å cube) at the 1969 Grenoble Conference on Electron Microscopy.

Baez and El-Sum took up Gabor's ideas for the electron microscope and examined the feasibility of applying them to X-ray microscopy. They concluded that to improve chances of success, they needed longer wavelength X-rays of narrower wavelength pass band, smaller source size, and a recording media of greater resolution.

Despite press rumours, an X-ray laser has not yet appeared and the two stringent requirements, outlined by Baez and El-Sum, cannot be satisfied in the most desirable way. Nevertheless, two patents on X-ray microscopes were applied for by North American Philips Company and IBM, in the period 1965—1966. Both patents trace their source of inspiration back to the COR discovery of the Bormann Effect in 1950.

The Bormann Effect is concerned with the anomalous transmission of X-rays along the atomic planes of a thick crystal set for Laue diffraction. When an X-ray is incident (at the Bragg angle) on a perfect single crystal with parallel faces, two diffracted beams appear at the outgoing face. One beam is diffracted into the forward direction and the other beam is diffracted downwards at twice the Bragg angle with respect to the forward direction. The latter beam has anomalously low absorption. Moreover, the diffracted beams have spatial coherence because they emerge from the back face of the crystal at the same point with phase-coherence. Thus an X-ray beam-splitter is available for the construction of the interferometric part of a two-stage X-ray microscope.

The ideas in the IBM patent are apparently not new. Indeed, it could be that the survey article by Möllenstedt (1966) proves some publication prior to the date of application of the IBM Patent[1] (this is certainly true of the IBM Crown Patent). Furthermore the pioneering work of Bonse and Hart in the 1960's utilized the Bormann Effect and a novel analyser crystal, which transforms the atomic scale fringe pattern into a macroscopic pattern recordable as a hologram on photographic film. Note that both patent specifications have a great similarity to the Bonse—Hart work.

Michael Hart and Ulrich Bonse received the American Crystallographic Association's Warren Award for their work in August 1970. Hart lectures at the University of Bristol and spent some time on sabbatical leave as a U.S. National Academy of Sciences Senior Research Associate at the NASA Center, Cambridge, Mass. Bonse holds a chair in physics at the University of Munster.

Feasibility of an X-ray Holographic Microscope

It should be clearly understood at the onset of this section that the author does not claim any detailed expertise[2] in X-ray physics

---

[1] Although Bell Telephone Laboratories and IBM Corporation were both supporting distinguished theoretical work by Cole and Batterman on the dynamical theory of X-ray diffraction, before July 1964 (see *Rev. Mod. Phys.*, 36, No. 3, (1964)).

[2] As a physicist, I would not normally proceed in the present manner (without submitting the views expressed to referees). However, I wish to demonstrate the difficulties faced by analysts when they attempt broad vista problems of evaluating and assessing COR. Without doubt, the major stumbling block for analysts is that they frequently have to analyse subjects well outside their own range of speciality. Notwithstanding the above remarks, it is still necessary to apologize to the reader who is a generalist for the included technicalities which will make this chapter slow reading.

and the views expressed on technical feasibility have not been subjected to extensive scientific comment.

X-ray microscopy by reconstructed wavefronts without Bormann Effect or Moire Effect in analyser crystals, suffers the limitations outlined by Baez. Under these conditions the resolution limits of competing types of hologram are determined by the geometry of recording arrangements and the resolving power of film.

Stroke and Falconer have shown that a useful gain in the limit of resolution can be made by using "Fourier-transform holography" to overcome the limitations of "Fresnel-transform holography" (see Stroke, 1967). In Fourier-transform holography, the geometry of the recording system is such that the intensity recorded on the hologram contains the Fourier-transform of the transmittance (together with the conjugate of the transmittance) of the object under examination.

Normally, Fourier-transform processes require the use of lenses but, as we have seen earlier, the low refractivity of X-rays makes this impossible. The so-called technique of "lensless Fourier-transform holography" does away with the lens but requires a point reference source (i.e. a spherical wave) close to the object.

Stroke claims that with angles of X-ray diffraction as small as $10°$ and point source to object spacing of 0.1 mm, a 100-fold gain in resolution is obtained by using a Fourier-transform hologram.

To obtain an appreciation of first-order requirements for Fourier-transform holography, it is useful to consider the difficulties of making an X-ray hologram of a flat circular object (radius r). When the object is placed in an object plane at a distance d from the hologram plane and illuminated with a coherent beam of wavelength, λ, the resulting Fraunhoffer-diffraction pattern [1] obtained on the hologram recording plane is a central spot surrounded by a series of concentric rings, the intensity of which is given by a Bessel Function of first order divided by its argument.

The size of the central spot is determined by the zero of the Bessel Function and its radius is given by

$$\frac{2\pi r}{\lambda d} \text{ (radius)} = 3.84$$

thus

---

[1] Fraunhoffer-diffraction patterns are obtained when the object to hologram distance is very much larger than the square of the object radius divided by the wavelength (i.e. $d \gg r^2/\lambda$).

Radius of diffraction spot $= 1.22 \dfrac{\lambda d}{2r}$

If the centre of the object and the point source (both lying in the object plane) are separated by a distance Δ, the resulting first-fringe width is given by $(\pi/2)/(2\pi \text{ frequency})$, i.e.

Fringe width $= \dfrac{\lambda d}{2}$

If the hologram is to record sufficiently a diffraction pattern, say N times larger than the spot size, we must have many fringes (to carry information) per detail of diffraction pattern, i.e.

$$N\, 1.22 \frac{\lambda d}{2r} \gg \frac{\lambda d}{2}$$

from which is obtained a restrictive condition on object size.

$$\frac{r}{1.22N} < \Delta$$

Note that since N will be greater than unity, the condition is sufficiently strong to avoid geometrical overlap between reference and object.

Another elementary requirement is that the film resolution (say, L lines cm^{-1}) should be larger than the fringe frequency, i.e.

$$\frac{\Delta}{2\lambda d} < L$$

A combination of the last two inequalities yields

$$\frac{r}{1.22N} < \Delta < 2\lambda dL$$

(See Winthrop and Worthington (1965) for greater detail.)

The Fraunhoffer conditions imposed on the holographic configuration ensure that the intensity recorded on the hologram contains the Fourier-transform of the transmittance without complicating terms of magnitude proportional to the exponential of $\pi(\Delta^2 + r^2)^{1/2}/\lambda d$.

In the absence of an X-ray laser, experimentalists will, no doubt, attempt to obtain larger coherence lengths from conventional sources; i.e. they will attempt to increase $(\lambda^2/4d\lambda)$ where $d\lambda$ is the wavelength bandwidth. A Cosslett—Nixon point-focus tube with metal target could be used as proposed by Baez and El-Sum, but even then the coherence length would be small. For CuK_α radiation $(\lambda^2/rd\lambda) \sim (1.5 \text{ Å})^2/4.5 \times 10^{-3} \text{ Å} \sim 0.1 \times 10^3 \text{ Å} = 10^{-6}$cm!

The extremely short coherence length determines the path differences over which temporal coherence can be maintained and predicates the need for small-sized holographic configurations. These, in turn, through the crude inequalities derived above, predicate the need for small object-to-reference source distances and smaller object sizes. If allowance is made for the effect of finite size of reference—source pinhole, the present outlook is, to say the least, challenging.

A theoretical estimate by Winthrop and Worthington (1965) (made for U.S. Public Health Service), of a Fourier-transform holography arrangement is: X-rays of 1 Å wavelength, object size less than 2 μm, object—reference spacing 3 μm predicts resolution of 500 Å with a pinhole of 0.05 μm diameter. The small sizes of pinhole and object in this requirement are severe limitations and imply large exposure times with mechanical stability problems to be overcome.

X-ray microscopy by reconstructed wavefronts, incorporating Borman Effect wave splitters and Moire Effect analyser crystals, appears to offer two immediate advantages over the kind of microscopy described above. The Bormann Effect is useful in overcoming difficulties of achieving spatial coherence and the Moire Effect offers a means of overcoming the present limitations of resolution of photographic film. (See the IBM Patent.) Using Cu K_α radiation, Borse and Hart have observed dislocations in crystals over areas of five millimetres diameter. (The dislocations were introduced as elastic strain by evaporating aluminium film discs (50 Å thick) onto the silicon analyser crystal.) But such experiments are only partial confirmation of the validity of the methods described in the patents, for example the Borse—Hart experiments do not confirm that it is possible to introduce uniform strain into the analyser crystal as described by IBM.

Photographic set-ups, using Borse—Hart interferometric techniques, still suffer limitations due to lack of temporal coherence (small coherence length) even when "filtered" X-radiation is used. Two conflicting opinions are held by experts on this last point: it is thought, in one University laboratory, that an X-ray laser is required, but some ex-Mintech laboratory staff believe that there is some hope of achieving X-ray holography without a laser by using a Lloyd's mirror arrangement. Both laboratories suggest that the limit on resolution will not be better than 100 Å.

Obviously there are some good ideas on the subject of X-ray microscopes but people are understandably taciturn where their ideas are speculative. If a good "reflection" lens could be manufactured, one of the difficulties of conventional Fourier-transform holography could be overcome by using lens-focussed X-rays instead of a pinhole for the reference beam.

At this stage of progress in X-ray COR work it would be foolish to predict lack of success within five years of either kind of holographic X-ray microscope; even with present technology there is a finite probability of success. If new methods of recording fine fringe spacings (photoelectric?) are developed and/or X-ray lasers become a reality, one would predict almost certain success and be tempted to lower the current estimate of resolution limit.

It is thought that some feasibility work has been carried out on gas X-ray lasers in the U.S.A. (although during this assessment this was not tracked down). Rumour suggests that, since normal incidence end-reflectors at X-ray wavelengths are impossible, gas-pumping currents of 20,000 Å are required. It is possible that plasma physics will also yield large soft X-ray yields from plasma focus or synchrotron radiation devices[1].

Possible Benefits of X-ray Holography

There are obviously dangers and difficulties in attempting to canvass opinion on the likely uses for an holographic X-ray microscope not yet in the development stage. Even if it is possible to complete a market survey, without unduly raising the hopes of hard-pressed microscopists and crystallographers, socially orientated arbiters could not be persuaded to support a COR investment decision based on purely economic grounds. For example, the Nobel Laureate in Physics, Professor Dennis Gabor has commented upon the possibility of a cure for cancer and other pernicious diseases and suggested that "a thorough scrutiny be made for all physical methods which have not been applied to cancer re-

---

[1] After completion of this manuscript it came to my notice that Kepros, Eyring and Cage claim to have generated coherent X-rays (see *Proc. Natl. Acad. Sci.*, 69, 1744 and "The X-ray Laser makes its Debut", *New Scientist*, 3rd. Aug., 1972).

search (examples: microspectrography of cells by Fourier spectrograph and by holography methods)". Commendable lobbying has also been made for more support of molecular biology and structures like the European Molecular Biology Organization (EMBO) council has claimed that their proposals for a central laboratory "would provide a number of services to the European biological community at large, *such as pioneering in the fields of high-powered instrumentation......*" (my emphasis).

Such appeals, based on a chimera of possible social benefit, cannot by lightly dismissed, even at this stage of ignorance about the precise value of the likely social benefit, because COR is essentially a speculative venture and the relative amount of high-risk capital at stake in COR on molecular structure may be inferred from the following "league" table.

Normalized cost of producing an increment of knowledge in various fields of COR[1]

Elementary particle physics	1.00
Astrophysics	0.55
Nuclear physics	0.26
Solid state physics	0.25
Atomic and molecular physics	0.10
Chemistry	0.04

The table would indicate that a discovery in molecular physics is relatively less expensive than some discovery in other fields.

Orders of magnitude of benefit to be obtained from the market in X-ray diffraction equipment may also be inferred by anyone brave enough to predict the market penetration or induced demand likely to be achieved by holographic microscopes, from the following statistics[2].

World Market for X-ray Diffraction Apparatus

| 1965 | £2.5 to £3.3 million at £8,300 each |
| 1968 | £3.7 to £4.9 million at £12,000 each |

[1] I have taken some liberties here with a table produced by Harvey Brookes in *Science Policy and the University*, published by The Brookings Institute in 1968.

[2] Both statistics taken from OECD's useful work "Gaps in Technology—Scientific Instruments".

Bio-medical instruments (X-ray equipment)

	Europe	U.S.A.
1966	?	£42 million
1967	£19 million	£48 million

User benefits of such a speculative instrument as an X-ray holographic microscope would critically depend upon the finesse achieved by the instrument. *If* the instrument had adequate resolution to assist in determining three-dimensional structure of proteins, nucleic acids, and enzymes it could be assumed that the user benefits would almost equal the yearly opportunity cost of allowing highly talented scientists to pursue the present painstaking, "empiric", methods of structural determination.

There are, of course, many potential areas of benefit for X-ray microscopes in markets which have been opened up by electron microscopy. The penetration depth of X-rays is still fundamentally large compared with that of charged particles (electrons). As pointed out by Sugata *et al.* (1970), theoretical curves of (observable specimen thickness) versus (accelerating voltages) are curves of diminishing returns. Observable thickness is predicted as 1 unit at 100 keV, 2.4 units at 500 keV, 2.9 units at 1,000 keV, and 3.2 units at 3,000 keV but there is some experimental evidence that these theoretical estimates are pessimistic. There is, therefore, a price to pay for progress in analysing thicker specimens with an electron microscope. Indeed, Hitachi Ltd. have revealed that production costs increase linearly with voltage, i.e.

Production cost $\sim$ 0.13 (KeV) million yen
$\sim$ £0.15 (KeV) thousand

Thus, in order to obtain penetration of specimens of biological material of 3.0 μm thickness, one would need to spend five to fifteen times more on a higher voltage electron microscope than on a 100 keV microscope which could manage a specimen of limiting thickness of one micrometre. Alternatively, an X-ray microscope, operating at a wavelength of one angstrom would give reasonable penetration depths of 20—50 μm with some differentiation between different atomic constituencies because X-ray penetration is proportional to the cube of atomic number.

An estimate of benefit accrued by avoiding the need for very high voltage electron microscopes in examining thicker specimens is not currently feasible because X-ray microscopes (which use holography) are not currently developed or priced. The benefit could conceivably exist in the future if the use of higher-voltage electron microscopes becomes essential; doubly so if the higher-energied electrons tend to destroy the integrity of the observed molecules.

Possible Links Between Current Research and a Normative COR Goal

At the termination stage of this prospective study, a positive recommendation on constructive methods of organizing effort towards a normative goal was not possible. Indeed, to make such a recommendation based on the information gleaned would have been presumptuous. All that could usefully be done was to comment upon specific parts of the Mintech's research programme.

Initially, some thirty recognized experts on crystallography, biochemistry, biophysics, microscopy, and pathology were all canvassed. Only five of the people approached yielded information relevant to the prospective study. From the remainder, various responses were obtained: some gave encouragement for attempting such a speculative study but, regrettably, others handed out rebuffs which could only be interpreted to mean that they did not want non-specialists to comment on their field. Centres of excellence, like the Cavendish Laboratory Cambridge and the National Physics Laboratory, undoubtedly showed the greatest tolerance to what initially must have seemed (and may still seem) shallow questions. Furthermore, these organizations conclusively demonstrated their cogent background knowledge and this showed that the possibility of producing a viable high-resolution X-ray microscope had never been far from their minds[1]. Both centres have a successful commercial record in the development of instrumentation arising from COR which suggests that if a breakthrough in X-ray microscopy occurs they will at least know how to make the first-order assessment of its commercial value. (This does not necessarily mean that the centres would wish to pursue the commercial exploitation themselves.)

The personal recommendations made by this author naturally

---

[1] Theoretical "justification" of centres of excellence suddenly came into practical perspective when I was faced with obtaining information quickly.

282

put a great deal of reliance upon the centres of excellence. However, since I made my suggestions directly to Mintech staff who are concerned with their own in-house programmes, it is not fitting that my subjective advice be repeated here. Suffice it to say that the actual research already sponsored was second-to-none and assessment of its central management and funding can be personally biased.

Concluding Comments on the Prospective Evaluation of COR

Difficulties of attempting an evaluation of COR have been described early on in this chapter. My own reaction to the study is that two principle questions arise.

"Is there any point in trying to assess the future general outcome of COR when it can be argued that all attempts to acquire knowledge are a justifiable end in themselves?", and

"If there is a need for this kind of technology assessment, and if we accept the inherent difficulties outlined earlier, how can the process be efficiently conducted?".

There are two main answers to the first question. Some respondents would say "no" because they claim that the direction of COR will be determined by the propensities of the researchers and the fashion of the moment (which may be as unpredictable as the fashion in apparel). Others would say "yes" because, in their blacker moments, they see the COR scene as mainly comprised of researchers galloping "like hell along the wrong racetrack" and they would wish to see some societal control effectively determining the general direction of the COR activity. Whatever answer is preferred, there is a need for *more* information about COR activities and, if one believes (as I do) that the second answer is correct, the need is based upon wider consideration than that of simply enhancing the efficiency of the COR that just happens to be currently in vogue.

Apart from military intelligence systems, there is not a great deal of organized effort devoted to the intelligent collation of formation. Each COR unit, in these days of fashionable data banks, has some sort of electronic data processing aid (if only a system operated by a library to keep surveillance over the ever-increasing volume of publications). Yet the *real* intelligence system of COR units is based upon personal contact and private communication between specialists. By contrast, central management in

283

government, university, or private organization has to rely on their own COR workers or a distinguished promotee from the COR ranks. Such promotees, as "chief advisors", usually recruit staff to assist them with their function of proferring advice to the arbiters-of-decision. The recruited functionaries are not usually the people who possess research insight. Normally they have other dominant qualities which they sensibly wish to use. It is very rare indeed to find a man possessing insight into research who will take such an administrative post and align himself with central management. *The catalytic men who could help to evaluate COR most effectively tend to be in COR organizations, with group loyalties and a tendency to be affected by colleague enthusiasms.* Therefore, in my opinion, the answer to the second question is "Yes. Persuade or coerce a non-negligible proportion of the catalytic men to leave the laboratories and pursue the specialized aspect of technology assessment which is concerned with the prospective evaluation of COR".

Some way must be found of breaking the correlation that seems to exist between research-insight and an unwillingness to be employed on anything other than the actual research activity. No amount of systematized information processing capacity will ever compensate for the lack of catalytic men or the fact that management-orientated staff (possessing no great penchant for COR but with other indefinable compensatory attributes) seek so successfully to align themselves with central decision making on COR.

The recent pause[1] to the *growth* of funding for all research should have given the catalytic men some incentive to seek equally satisfying employment as evaluators and assessors, if only to reap the reward of helping achieve a sensible allocation of the available

---

[1] The quotient (RD/GNP) for the United States fell almost linearly from 3.04% in 1964 to 2.79% in 1970 and other countries tend to follow the American precursor. The annual rate of growth in real terms of the science budget now administered by the U.K. Department of Education and Science has fallen from 12% in the mid sixties to 4.6% in 1971 and will probably fall still further.

"Well, here we are in 1970 after these twenty years and it is evident to everyone that exponential growth of expenditure on science has indeed declined." "At the most optimistic interpretation, the compound interest rate of growth has been changed to simple interest; that we shall approach saturation in wild and wasteful oscillatory swings below and above the reasonable marks as one side wins the argument and then the other".

Derek J. de Solla Price (1970)
Avalon Prof. of History of Science, Yale.

funds to areas of COR showing greatest "promise". There has been no lack of recruits to the assessment function from applied research and development, yet the reaction from good COR men has been to move from poorly funded research to better financed research projects[1]! Catalytic COR men have always voluntarily served on advisory committees providing that the committee work does not take too much of their research time. Service of this kind gives some protection against any tendency of others "wrongly" to evaluate COR activities but it hardly provides the kind of sustained appraisal necessary for the selection of a "healthy" portfolio of investment in COR. Unless some catalytic men move into the assessment sphere, the solace obtained from denigrating those who would elect themselves to the position of assessor on the strength of "intangible" managerial ability, will be poor compensation for potentially irreversible harm to COR. (Verbal tilting at a quintain is no fun when the critic eventually receives as heavy a blow as his target.) Dispassionate contributions to the techniques for the evaluation and assessment of COR are needed urgently before it is too late and COR men awake to find themselves the chattels of business school graduates and other acolytes of management.

Accurate statistics showing the ratio of competent impartial assessors to researchers are not presently available and no useful purpose is served by making inferences from published figures of expenditure on "administration" and "research". Some feeling for the ratio may be obtained from the fact that the PAU, which caters partially for the assessment needs of the UKAEA and DTI, has a working strength which very rarely exceeds more than thirty professionals and they assess projects upon which thousands work.

Allowing for temporary secondment of staff to the assessment function, I would suggest that, in general, the ratio could be as low as one in five hundred. What is fairly certain is that in government-sponsored research there is not even enough assessment effort to keep accurate check on current prospects for on-going research projects which have already been sanctioned.

Methods of redressing this imbalance are not difficult to conceive. Proper career structuring of scientists would ensure that they spent some time seconded to the assessment function to

---

[1] The catalytic COR man is not so attracted to higher salary as he is to interesting and properly funded research, though, ideally, he would like both.

assist with the assessment of science in a field where they have not got laboratory loyalties. Further, it should be de rigueur for preferment to directing posts, within COR organizations, that the potential leader should not only have spent, say, two years discharging the assessment function, in addition he should have proved objective and sound in his judgement[1].

References

Booth, A.D. (1968). *Fourier Technique in Organic Structure Analysis*, Cambridge University Press.

Brothwell, D. (1969). *Science in Archaeology*, Bristol: Thames and Hudson.

Byatt, I.R.C. and Cohen, A.V. (1969). "An Attempt to Quantify the Economic Benefits of Scientific Research", *Department of Education and Science, Science Policy Studies No. 4*, London: Her Majesty's Stationery Office.

Cosslett, V.E. and Nixon, W.C. (1960). *X-ray Microscopy*, Cambridge University Press.

De Solla Price, D. (1970). Hearings before the Subcommittee on Science, Research, and Development of the Committee on Science and Astronautics, U.S. House of Representatives, p. 649, 91 Congress, 2 Session, July—September, Washington, D.C.: U.S. Govt. Printing Office.

Gabor, D. (1949). "Microscopy by reconstructed wavefronts", *Proc. Roy. Soc.*, A 197.

Gruber, W.H. and Marguis, D.G. (1969). *Factors in the Transfer of Technology*, pp. 155—176, Cambridge, Mass.: MIT Press.

Hosemann, P. and Bagchi, S.N. (1962). *Direct Analysis of Diffraction of Matter*, Amsterdam: North-Holland.

IBM (1969). *Brit. Patent 1,147,150.*

Jerrard, H.G. (1969). *New Scientst*, 10th. July.

Langrish, J. *et al.* (1972). *Wealth from Knowledge. A Study of Innovation in Industry*, London: MacMillan Press.

Möllenstedt, G. and Wahl, H. (1968). *Naturwissenschaften*, 55, 340.

Möllenstedt, G. (1966). *Conference on X-ray Optics and Microanalysis*, p. 15.

New Scientist (1969). *New Scientist*, 21st August, 370.

SAPPHO (1972). *A Study of Patterns of Innovation in the Chemical and Scientific Instrument Industries*, a report delivered to the Science Research Council by the Science Policy Research Unit (SPRU), University of Sussex.

Stroke, G.W. (1967). *An Introduction to Coherent Optics and Holography*, New York and London: Academic Press.

Sugata, E. *et al.* (1970). "Project for Construction and Application of a 3 MeV Electron Microscope", *Microscopie Électronique 1970*, 121.

---

[1] I make this point to offset the probability that a period spent acquiring relatively superficial expertise as an assessor could add to the plausibility of a "robber baron" of science when he is a postulant for funds.

286

Tomita, Hiroshi (1970). "Electron Microhalography by Two-beam Method", *Microscopie Électronique 1970*, 151.

TRACES (1968). Technology in Retrospect and Critical Effects in Science 'TRACES' ", prepared for the National Science Foundation by the Illinois Institute of Technology Research, Vol. 1, 1968.

Winthrop, J.T. and Worthington, C.R. (1965). "X-ray Microscopy by Successive Fourier Transformation, *Phys. Letters*, 15, No. 2, 15th March.

CHAPTER 9

Ethical Industrial Intelligence

"Intelligence deals with all the things which should be known in advance of inititating a course of action."

Dulles (1969)

Introduction

The above quotation might have been appropriate when A. Dulles was head of America's CIA and running one of the largest covert intelligence organizations the world has ever known. In technology assessment, however, it is impossible to know *all* things (in advance of completing an assessment) which would influence those who determine preferred courses of action. The best that can be done is to exercise some discrimination in determining what information can, and should, be acquired during analysis; therefore, some modification of Dulles's definition is appropriate.

"Ethical industrial intelligence deals with the attainable information which should be known in advance of completing a technology assessment."

Throughout this chapter the above definition will be adhered to and discussion will concentrate on what is and is not ethical and how information can be obtained and collated.

The doyen of Europeans who have recently shown interest, outside covert organizations, in industrial intelligence is Dr. Frank Pearce. Pearce's publication *"Intelligence: A Technology for the*

1980's?" has created sufficient interest for him to begin organizing Europe's first International Conference on Intelligence (see Pearce, 1971 and 1972). Although this chapter owes much to my own contact with Dr. Pearce, nevertheless it should be understood that the following comment is solely my own.

Ethical Intelligence

There is no clear-cut law of morals which yields an unambiguous definition of what the professional code of conduct should be in the acquisition of information for technology assessment. Most analysts are convinced that they should not stoop to espionage, which is defined by Pearce (1971) as

"the means by which information is acquired for purposes not in the interests of the rightful owner".

Nonetheless, the "interests" of the rightful owner are variable (depending upon what he takes to be the operating environment in which his interests arise) and "rightful" ownership of something as intangible as an idea has always been notoriously difficult to define.

The following anecdote, taken from my own experience, might illustrate some of the difficulties faced by analysts in deciding "what is ethical?". World-wide interviews were conducted, during a technology assessment, with technologists experienced in a particular field of high technology. One specific piece of hardware which had been designed by the leading technologists was understandably not available for physical inspection. Nevertheless, the foreigners in the van of progress were very generous in allowing some discussion of their technology. Interviewers were shown various plans and sketches and were allowed to question the experts, who were interested in generating market interest[1].

The interview terminated with the experts giving their interviewers plastic scale-models of their technological hardware. After ascertaining that the plastic models contained no deliberate defects, the interviewers were able to use them in an analysis aimed

---

[1] It is useful always to remember that, although one may seem to be the only recipient of information, even the content of questions asked reveals much to the interviewee about the state of one's own technology.

at determining the performance characteristics of the real hardware.

Various inferences and deductions can be made from the above anecdote.

(1) The undoubted advantage of obtaining the model could never have been gained without the analysts being willing and able to travel abroad to seek information by personal interview.

(2) Had the model not be offered and had it been stolen, the thieves would have been embarking upon the first step in espionage. (In fact the admiration expressed by the interviewers, after *seeing* the model openly on display on the interviewee's desk, was sufficient to elicit the impromptu offer of the model.)

(3) The interviewees certainly knew of the possible (although not probable) damage to their own interests if the interviewers recommended that their organization build a facsimile of the leading technology[1]. In other circumstances the interviewees need not necessarily have been so guarded and then their own interests would have been at risk. The point being that the interviewers could have claimed that the models had fairly become their property and any information consequently deduced from them not already covered by patent and copyright would be as much their property as of the original owners of the model.

(4) Perhaps the greatest danger to donors of information is that they judge the value of the information imparted against what they know at the time of transfer. When the recipients of information have greater insight about possible areas of advantage to themselves *and* the donor but do not possess the know-how necessary to bring the potential advantage to fruition, there is great disadvantage to the donor of information. (A corollary is that if the interviewer can make the donors of information think he has this insight then, obviously, the donors may be more generous.)

After surviving many interviews similar to that described above without creating undue offence, and observing the successful behaviour of past colleagues, my own guidelines for behaviour in similar situations are as follows.

(1) When an interview has been granted *after* the interviewee has examined the interviewer's stated reasons for requesting discourse, the interviewer is under no further obligation to become

---

[1] In fact these interviewees were so experienced and astute that, before consenting to an interview, they almost certainly weighed possible advantage and disadvantage to be gained from the interview.

expansive about his own intentions. (However, in practice, the interviewer may be forced to become more informative in order to engender freer exchange of information during the interview.)

(2) The interviewer may attempt any form of overt copying (or recording) *provided* he does so with the interviewee's permission.

(3) Under no circumstances must the interviewer make covert records or photographs on the interviewee's premises [1].

(4) The interviewer may attempt to reconstruct information (plans, equations and data) he has seen during the interview but only *after* he has left the interviewee's premises.

(5) The interviewer is under no obligation to reveal that the interviewee has failed to perceive an area of potential benefit.

Each assessor of technology will naturally make up his own mind about what unethical intelligence activity is, unless his terms of employment or membership of a professional organization have committed him to an agreed code of conduct [2]. My own general position is that, in cases where organizations do not wish to reveal information which is not covered by the Official Secrets Act or any other legislation, they present a worthy challenge to the ethical intelligencer and it is permissible to try to obtain the information by any means that are not underhand. (A classical example of this, always quoted to market analysts, is where producers are not willing to reveal their production figures which can be obtained ethically by counting units of transportation to the producers outlets.)

In reaching a personal code, it is useful to remember that laws usually penalize most severely those actions which lead to the greatest damage and the greatest amount of technological transfer (and hence information leakage) can be obtained by suborning, or moving people [3].

Therefore, it is not unreasonable to suggest that each assessor has the following necessary clause in his code of ethical conduct.

---

[1] The nuances of covert copying are not properly covered by law. However, most analysts would agree that the use of a hidden camera or tape-recorder is unethical.

[2] Reputable organizations, e.g. market research associations, will recommend a code to their membership. Government organizations are particularly careful and, to my knowledge, one department would not even allow the open use of tape-recorders (even with the permission of the interviewee) because of the possibility of future embarrassment.

[3] For a discussion pertinent to this point, see Langrish *et al.* (1972), p. 10.

> Thou shalt not suborn another organization's employee for the purpose of obtaining industrial intelligence.

Technology itself intrinsically enters the problems of defining a clear demarcation between ethical intelligence and espionage because the rapid development of new techniques for handling information multiplies possibilities for obtaining (overtly and covertly) information. Subjective comments on possible ethical codes of behaviour therefore become rapidly dated. Current suggestions in the literature include Hamilton (1967) which contains the Wade system of gradation. Whatever the impact of future technology of information will be, however, a simple deterrent would be that "unethical collection of information does not pay."

Having been brought up, not entirely satisfactorily, to believe that virtue has its reward, my own prediliction towards ethical intelligence behaviour is based upon the belief that

> the useful set of information sources available to the analyst, discovered to be covert or unethical in his collection of information, ultimately shrinks with time, thus forcing unethical intelligencers to greater and greater clandestine efforts, and expense, if they wish to maintain their information flow.

Perhaps some future researcher will amuse us all by discovering some "sod's law" of unethical intelligence which accounts for the escalating expenditure of the world's clandestine intelligence agencies. In the absence of such a "moral" discovery I suggest that a conceptual model of intelligence behaviour should be somewhat as follows.

The rate of acquiring information from a particular sub-set of information sources is a monotonically increasing function of

(a) intelligencer's ability (A),

(b) intelligencer's expenditure ($£$), and

(c) what the source of information *thinks* the intelligencer's ethical rating is divided by the intelligencer's *actual* ethical rating (E_0/E).

If it is agreed that any limited sub-set of information sources has only a maximum amount of information available (say i_{max}), then a first-order model for the growth of intelligencer's acquired information (say, i) would be

$$\text{Intelligencer's information} = i = \frac{i_{max}}{1 + C_1 \exp(-A.£.E_0.t/E)}$$

Which is a logistic function (see Chapter 7) with t representing time and C_1 is a constant determined by how much intelligence about the sub-set the intelligencer begins with.

On the other hand, the faster the rate of acquisition of intelligence by clandestine means, the greater one would expect the probability of discovery to be, i.e. the probability is increased that the sub-set investigated realizes that the intelligencer's apparent ethical rating is too high. Another hypothetical logistic function which might represent this effect is

Probability that the
intelligencer's real ethical $= p = [1 + C_2 \exp(-E_0 t/E)]^{-1}$
rating will be discovered

Where C_2 is a constant propensity to suspect that things are not as they seem.

After delving into a particular sub-set of information sources, the intelligencer will have gained information but had his real ethics exposed, through p. The amount of information actually gained will depend upon the time spent acquiring it and this time ultimately yields diminishing returns because both the above functions are concave towards the time axis, after half the available information has been obtained. Whatever happens, at the end of the intelligence activity in the sub-set there will be explicit amounts of information gained and suspicion created. The suspicion created (measured by p) will not unreasonably determine the vigour with which the sub-set communicates its suspicions to the other information sub-sets. Depending upon the speed of communication between sub-sets and the degree of suspicion communicated, the intelligencer will begin his next assignment with a reduced ratio of apparent to actual ethical rating (i.e. (E_0/E) will decrease).

Thus a new apparent ethical rating quotient will be used by other sources of information subsequently approached by the beady-eyed unethical intelligencer. Conveniently, this new quotient may be put into inverted logistical form as follows.

$$\frac{\text{Apparent ethical behaviour}}{\text{Actual ethical behaviour}} = \frac{E_0}{E} = 1 + C_3 \exp(-tC_4 p)$$

where p is determined by the penultimate equation, C_4 is a constant determined by the maximum speed of communication between sources of information, and C_3 is a constant determined by the ratio of apparent to actual ethical behaviour generally assumed before denigration began (i.e. $C_3 = (E_0/E) - 1)_{t=0}$). Note that it is important to realize here that the last hypothetical logistic function is inverted to allow for the fact that denigration of the unethical intelligencer leads to a less favourable ratio of (E_0/E) (i.e. with increasing denigration this ratio must decrease towards unity).

The above three equations may determine the hypothetical cycle in the decline of the untrustworthy intelligencer. The last equation gives the value of ethical rating-quotient which must be substituted back into the first two equations when the intelligencer recommences a cycle of investigation. Because the quotient of apparent to actual ethical behaviour reduces with increased activity, the consequence is that if the intelligencer wishes to maintain his efficiency in collection of information he must increase the quotient by becoming even more unethical (i.e. he reduces the actual ethical rating E). This means that he is even more likely to be detected as untrustworthy (and indeed he has become more so) because probability of unmasking will grow more rapidly with time. Alternatively, the intelligencer can decide to behave as branded and work at his previous level of unethical rating, in which case the only way he can maintain his efficiency in collection of information (without increasing the denigration) is to spend more money (£ in the first equation).

The above conceptual cycle has not been numerically substantiated although it has the feeling of truth about it and it is well known that repeated acts of clandestine operation reduce effectiveness. My hope is that, in the face of increasing temptation to act unethically in the collection of industrial intelligence, the model will act as a sort of Hogarthian unethical intelligencer's progress with logistic tableaux instead of engravings pointing to dangers of embarking upon nefarious activities!

Intelligent Collection and Collation of Information

Before the advent of existing methods of electronic storage and retrieval of information (electronic data processing, EDP) and the construction of reliable high-speed data communication systems, only mechanical filing systems could be used to buttress or off-set the lack of finesse in intelligence activity. Today the information technology systems used are much faster and more reliable and

EDP provides some intelligencers with rapidly up-dated indexes of many categories of information which are invaluable in conducting ethical, neutral, offensive, or defensive intelligence activities. Some large companies, for example, possess EDP systems which will provide indices of science, technology, activities of potential competitors, world patents, etc. (see Chaumier, 1972).

The possession of such powerful EDP systems is, of course, no guarantee of the overall efficiency of any intelligence system because no amount of technological support can compensate for a basic lack of organization in the intelligence process. Organizations face the possibility of falling prey to the "degeneration syndrome" where, in the absence of sound ideas for the direction of intelligence, they degenerate to the analysis of statistics, in a general way, hoping for a blinding flash of significance to be revealed and in the absence of available statistics they degenerate, still further, into the relatively aimless collection of statistics which they can analyse. This degeneration syndrome is illustrated by the following iconic model.

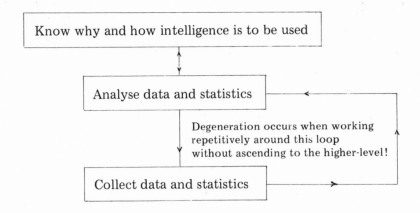

Organizations which have "degenerate" intelligencers can be made less efficient by possession of an EDP facility simply because the loop between collection and analysis of statistics becomes automated and its use becomes an end in itself for intelligencers who are unwilling, or unable, to define carefully *why* they are pursuing the degenerate activity.

Most EDP systems can do more than provide general-purpose information like that provided in circumscribed and defined fields by valuable organizations like Euratom, ESRO, and The Institute for Scientific Information. General information centres are extremely useful and should be utilized for the provision of abstraction services and general documentation. However, for the collection and dissemination of intelligence, really specialized EDP systems are required which will provide information that has been specially suited to the "intelligence profile" of the user's requirements. This "intelligence profile" in some systems is carefully defined by "key words" that are used to interrogate the stored, and continually up-dated, information. Obviously, such specialized EDP intelligence systems should be kept under careful and direct surveillance by users because a useful step for a competitor would be to obtain knowledge of the "key words" or "descriptors" defining information profiles. Since "profiles" are, of necessity, highly personal and revealing, it is not appropriate that they be discussed here, the reader who is unfamiliar with the use of such procedures for systematic search should read the non-confidential examples in Chaumier (1972).

In technology assessment, a serious omission in an analyst's "profile" can be very inhibiting to his analysis of future technological impact. It can happen that the profile defines the assessor's interest solely in technological terms without including the associated function-orientated uses of technology which happen to be of foreseeable interest. The analyst need not be, for instance, exclusively concerned with uses of methadone in controlling heroin addiction (see Etzioni and Remp, 1972). He might really be interested in *all* technologies which could help heroin addicts become acceptable members of society. (This last point is so obvious that I hesitate to make it but the difficulties of re-orientating the "mechanistic" parts of an intelligence system, once the original specifications have been made, seem to deter many from widening their computerized profiles.) Jantsch (1967) contains a relevant, precomputer age, anecdote. He reports Zwicky as suggesting that Professor Lindemann's obsession with solid fuel led him to discount the possible success of the German effort on war-time rocket technology [1].

---

[1] Zwick's morphological method of analysis (see Chapter 7) has much to recommend it in the definition of function-orientated technology. But alas, it is also subject to a degeneration syndrome of its own which can multiply problems and devour the time available to analysts. Bridgewater (1969) shows how the computer can be used once the general morphology of technological systems has been defined.

Notwithstanding the possible dangers inherent in the use of EDP systems in too loosely defined an intelligence or information system, there is an increasing use of such systems by administrations in the public and private sectors. The outcome of such a trend is open to speculation, my own opinion is that the differential in competence between larger companies and departments and their smaller competitors will be increased. For instance, multinational companies with their greater expertise in organization and wider communication-links, stand to benefit more from prompter and more accurate retrieval of organized information than the smaller companies. (And they can afford the equipment by virtue of economy of scale.) In the interstices between the larger groups who benefit in ethical ways, there will be smaller, less ethical "tick birds" of society selling personal information on such things as financial and criminal records. The spur to a lot of this activity will be the greater competitiveness in industrial markets. A propos of this Pearce (1971) tells us that

"it seems very doubtful that there has been much appreciation of the transferability of intelligence ideas. A marketing game is not all that different from a war game".

In such a climate, a concomitant requirement for assessors of technology will be that of access to modern intelligence technology which is not inferior to that possessed by the largest (and best?) industrial companies.

In Europe, the Organization for Economic Co-operation and Development (OECD) has perceived the likely problems associated with information technology and has published two papers on this subject: "Information for a Changing Society — Some Policy Considerations" and "Government Responsibilities in Information for Industry". In the United States a group of 37 chief executives from business education, public institutions, and foundations has sponsored (in co-operation with Science Policy Foundation, European Cultural Foundation) a comprehensive report entitled "Information Technology: Some Indications for Decision Makers", which is said to look "at the information environment for the Seventies and Eighties, speculates on its implications for major institutions and the individual, and identifies ten major problem areas and some of the initiatives in each of these areas which can, and should, be taken by leaders today". At the time of writing, the conference scheduled to discuss the above papers under the title "The Information Society of the 70's and 80's — A Transatlantic Assessment", has not yet taken place. However, individual organizations must design their own intelligence systems to suit their own idiosyncratic needs and, although presence at such a conference is a useful intelligence exercise in itself, individuals can start thinking *now* about their own intelligence requirements without the useful stimulus of an international jamboree. Notwithstanding the "barnum and binary" display of public concern, the intelligence function is one area where each entity is primarily on its own.

Attendance at international and other conferences is normally indispensable to the assessor of technology and provides an important intelligence input. Indeed, at any conference a behavioural scientist can note the continual formation, and re-formation, of active sub-groups — a sort of Brownian motion of seekers and collators of information which physically demonstrates the intelligence activity. An index which compares activity external with activity internal to the formal conference will tell the observer much about a person's penchant to gather information (if normal social intercourse is discounted). EDP cannot replace the input of information that is best gathered by personal contact, in fact EDP processes need this kind of input to operate effectively. Companies which restrict the number of associations their employees can join, whilst at the same time investing in EDP, take a short-sighted and unbalanced view of the intelligence process. Professional societies not only foster professionalism, they also act as clearing houses and "grapevines" for tenuous information which is likely to be more valuable than information which has been clarified (and possibly used) by many hands. A noticeable trend in Europe over the past few years has been for more and more public servants to join such organizations, not the older established institutions but the newer associations like those concerned with market research and "sciences" which have not yet become respectable[1].

Assessors of technology should consider membership of organizations which cater for many disciplines, and merge, say, industrialists and employees of executive agencies of government, to be

---

[1] The trend certainly exists and, depending upon the value placed upon it, it may be welcomed or not by the public service. It is remarkable that when the U.K. Ministry of Technology was at its apogee, the number of civil servants joining, for example, the European Association for Industrial Marketing Research was pitifully small compared with those who had been encouraged to become members of the major engineering institutes.

de rigueur. It is nearly as appalling to see narrow technologists concerned with the intelligence aspects of important assessments, as it is to see effete generalists in the same position. Technologists and generalists are both necessary to the intelligence function but neither should operate with their inputs of information circumscribed.

Impetus for the drive to introduce wholesale application of technology often comes from the following sources.

Inventors of new technology

Aficionados of technology

Entrepreneurs

Planners who see the need for specific technology

National companies

Multinational companies

Potential consumers

Governments

It is a truism to say that the order of this list, in descending order of importance and influence, varies from case to case; nevertheless, it is obligatory for the analyst to know what this order is before he makes his assessment. Without efficient collation of information from *all* sources, the assessor cannot rank the influences (and possible impacts) correctly and ultimately technology assessment cannot exclude strategic planning and ethical intelligence activities. A case is known to me where the correct ranking of the above list of influences, behind the possible application of a new all-embracing technology, enabled the analysts to put together the possible strategy of the most influencial group (a multinational company) and thus recommend a compatible strategy to their employers which included a necessary defensive part.

There is a prevalence of intelligence units in government, banking, investment companies, and industrial companies. We are even told by Pearce (1971) that a number of counter-espionage agencies exist in London alone and that a least two schools of industrial espionage are believed to exist, one in Europe, the other in the Far East. Yet there is no standard procedure taught on ethical intelligence. Military intelligence practice is sometimes aired in the literature and U.S. Brigadier General Platt has made the following check-list (see Platt, 1962).

(*1*) Clarify definitions and assumptions,

(*2*) extend known facts by deductive logic,

(*3*) use cross checks (cross-correlation of information available to an object which cannot be closely approached),

(*4*) search for significance. What is the mechanism of the organization examined?,

(*5*) use analogies and categories. What are the familiar/unfamiliar elements of the situation?,

(*6*) develop judgement.

Such check-lists are useful in a very limited way as the Pearce (1971) working group discovered but where every man is his own ultimate conscience of what ethical behaviour is, there are many difficulties in transferring military (and other partially covert processes) into the overt practice of technology assessment for ethical civil purposes.

Perhaps the best that can be done without offending decency is to develop an awareness of the importance of good contacts, correlative procedures, deductive logic, and being on the qui vive for unexpected developments from unexpected quarters. Perhaps we should encourage the gregarious collector of information who has discretion, discrimination, and an analytic mind rather than the "beady-eyed" introvert who will make a computerized mountain out of a molehill of mediocre statistics?

General Advice on the Organization of Intelligence in Technology Assessment

In keeping with the practice of the U.K. Civil Service, there was no organized covert intelligence activity in the Programmes Analysis Unit. But a perusal of Chapter 4 will show that the Unit has two important advantages of scale in the ethical collection of information. First, the list given in that chapter reveals an unusually wide variety of technology in the projects analysed by the Unit, and secondly, the numbers and variety of experienced analysts ensures a non-linear advantage in the generation of useful contacts. The variety of technology continually under review means that there is always a good chance that the Unit will become rapidly aware of any synergy which could arise between apparently unrelated technologies and n analysts yield not less than about $\frac{1}{2}n(n-1)$ exchanges of information within the group. (As admitted previously, this non-proportional enhancement of useful interaction has only been qualitatively experienced and its lower limit is only inferred from algebraic considerations.)

The second of the above advantages of scale suffers more than the first if an organization is organized into only partially communicating sub-units or "cells" (cellular organization has its advantages, we are told by the press and chroniclers of modern history, when protecting a covert organization from outside surveillance). With an overt organization there is no need to use one-cell teams of analysts unless there is an overriding need to preserve commercial security and avoid scaremongering before confirmation of real risk. An ethical intelligence infra-structure can, of course, have internal groups for the purposes of neat administration and to allow injection of control and leadership. But in the interests of efficient generation and use of information, there should be a maximum amount of, and if possible complete, communication between members of all groups across group boundaries. The suggestion is, therefore, that in order for ethical intelligence activity to prosper the team should be as large as possible within the bounds set by the need for the activity and the competence of management in avoiding the impersonal control that begets friction and consequent disaffection.

Parsimony over expenditure on travel and attendance at meetings and conferences is another inhibitor to good ethical intelligence activity. (Incidentally not a fault of the PAU, above mentioned.) Some organizations, in addition to paring the financial allowance, for these activities which are indispensable to the intelligencer, also make the bad mistake of allocating travel and conference allowances according to rank held within the organization when, to be logical, such allowances ought to be matched to staff's powers of deductive reasoning and their ability to elicit and judge what information is likely to be relevant. It has been known for organizations to lose most of their international contacts when a handful of staff have left the team. Therefore, it is good practice to

(1) Allocate outside contacts to the best interviewers but always ensure that they have understudies, and

(2) file visit reports together with all deductions and correlations between information gained on different visits.

The embarrassment caused by different interviewers from the same organization approaching the same contact with different official stories is too painful to contemplate and creates an impression of covert behaviour when an ethical course is being pursued badly.

Customary procedure, used to avoid confusion in a larger group, usually centres around a full-time "information", or "intelligence", officer. Unfortunately, however, it is also customary to delegate this duty to a staff member who is given status about equal to a junior in-house librarian and not much active encouragement, or incentive, to really exert himself in correlating documentary information culled from interviews, the Daily or the Technical press. If useful correlation is "nine-tenths of the game", some recognition should be made of the fact by allocating suitable status to the post of "intelligence" officer and asking creative staff members to serve short periods, in rotation, in this post.

Probably the best way of ensuring a "rarefaction of information" is for senior management to regard widening contact of their staff with outsiders as a danger to their own authority. Commonly, such behaviour on the part of senior managers is rationalized by saying "that there are always dangers in allowing too wide a number of contacts because staff not aligned-with-management may damage the organization's reputation". Nonetheless, no top management which "plays its cards close to its chest" can hope to have a truly efficient infra-structure of intelligent collection and collation of information.

> When a personal premium is attached to information by any member of an ethical intelligence group he should be moved elsewhere.
> When information is an essential working commodity it should not be hoarded by individuals.

This last advice is painfully obvious but it is remarkable how the difficulty of making good assessments, in the face of uncertainty, exacerbates normal behaviour and forces some people to retain information as personal insurance against possible failure to produce quality analysis. Enthusiasm and morale sustain the arduous search for relevant information and the surest way to destroy this sustenance is to give the assessors only partial information about the reasons why senior management require the assessment. (Close behaviour at Director-level begets "traitor's pools" and disaffected muttering in staff meeting places.)

> In the interests of efficiency the fullest possible reasons for requiring information should be given at the onset.

Even slight changes in the general background against which the analysis is conducted causes disproportionate reduction in the efficiency of the concomitant intelligence process.

> Repeat personal interviews with sources of information are most enervating when they are made necessary by inadequate initial briefing.

302

Concluding Comment

The present evidence is that concerted efforts to increase efficiency in the ethical collection of information are afoot in Europe. In a years time (mid-73), after the planned spate of conferences on intelligence, it should be possible to judge the likely impact of these developments on analysts in the public and private sectors who are dependent upon ethical intelligence activity. As was said at the beginning of this chapter, all the above comments are highly subjective and I have some temerity in putting them forward. But the strongest conviction I am left with, even after working with a Unit which was second to none in its ethical use of information, is that if technology assessment is going to be a useful societal tool we need much better ethical intelligence processes. The United Nations (1972) Conference on the Human Environment clearly recognises this need when it declares its intention to set up a "computerized catalogue" of sources of environmental information (see Chapter 13).

References

Bridgewater, A.V. (1969). "Morphological Methods — Principles and Practice", in: Arnfield, R.V., ed., *Technological Forecasting*, Edinburgh University Press, p. 211.

Chaumier, J. (1972). "Modern Methods of Documentation Offered to Marketing", *Industrial Marketing Management*, 1, No. 3, 309.

Etzioni, A. and Remp, R. (1972). "Technological 'Shortcuts' to Social Change", *Science*, 175, 31.

Hamilton, P. (1967). *Espionage and Subversion in an Industrial Society*, London: Hutchinson.

Jantsch, E. (1967). *Technological Forecasting in Perspective*, Paris: OECD.

Langrish, J. *et al.* (1972). *Wealth from Knowledge — A Study of Innovation in Industry*, London: MacMillan Press.

Pearce, F. (1971). "Intelligence: A Technology for the 1980's?", *Industrial Marketing Management*, 1, 11.

Pearce, F. (1972). in: Pearce, F., ed., *Proceedings of the 1st International Conference on Intelligence, University of Birmingham, 24th—27th July 1972*, Birmingham: Intelligence, in press.

Platt, Brig. Gen. (1962). "Guidance for Uncertain Evidence", *Military Rev.*, Dec. 1962.

CHAPTER 10

Universities and Technology Assessment

"Universities incline wits to sophistry and affection."
Francis Bacon

*"Planning and introducing a rational policy would imply the con-
scious development of roles for science and technology in the
pursuit of human and social purpose. The organization of science
and other forms of human thought towards a purpose may be
viewed in the framework of a multiechelon system built by empirical, pragmatic, normative, and purposive system levels. Structured approaches to science, education, and innovation may be
brought to congruence in such a system."*
Erich Jantsch [1]

Introduction

Widespread concern about present and future modes of interaction between technology and society has understandably led to
much re-thinking about universities. Protagonists of the "systems"
approach, like Erich Jantsch, have suggested that we attempt to
form new "normative inter-, and trans-disciplinary universities".
Other activists, like David Goldman of the University of Maryland,
have used existing universities to launch relevant courses on technology assessment. And international organizations, like OECD,
have worked towards and sponsored new international institutes:

---

[1] This quotation is from Jantsch (1972). (The emphasis is mine.)

the International Institute for the Management of Technology (IIMT), Milan, supported by the Austrian, British, Dutch, French, German, and Italian governments and some private industrial companies[1] and the proposed International Institute for Applied Systems Analysis (see Chapter 1).

In this chapter the part universities can play in improving the future assessment of technology and, increasing public-awareness about the importance of good technology assessments will be discussed. Throughout this chapter the popular expression "multidisciplinary approach" will be used, not to denote a general superficiality of approach spread randomly across boundaries separating established academic disciplines, but rather as a synonym for energetically concerted effort by a team of people comprised of competent experts from these academic disciplines which are relevant to the problem under attack.

Systems Universities

Before considering "systems universities" it is useful to have some understanding of the meaning attached to the word "system" by the new generation of "systems analysts" and experts in "cybernetics". According to Allport (1955) a system is

"any recognizable delimited aggregate of dynamic elements that are in some way interconnected and interdependent and that continue to operate together according to certain laws and in such a way as to produce some characteristic total effect. A system is something that is concerned with some kind of activity and preserves a kind of integration and unity; and a particular system can be recognized as distinct from other systems to which, however, it may be dynamically related. Systems may be complex, they may be made up of interdependent sub-systems, each of which, though less autonomous than the entire aggregate, is nevertheless fairly distinguishable in operation."

---

[1] Jantsch has unfairly pre-empted the intentions of the Governing Board and Director General of IIMT when he wrote in Jantsch (1972): "On the other hand, the grandiose idea for an international post-graduate 'systems university' to be located in Europe — a concept developed by an international committee and subsequently partly through OECD — collapsed because of lack of imagination in the moment when governments, and through them industrial confederations, became involved. The resulting .... (IIMT) at Milan (Italy) has now been set up by European governments on the basis of providing a framework for six-week training courses for industrial and public managers ....". IIMT is expected to do rather more than that, see Hill (1972) and Cade (1971).

Systems analysis is obviously the analytical attempt to understand "systems" as defined above and a "systems university", of the kind favoured by Jantsch, would address itself to learning about systems and systems analysis.

Experienced problem solvers and conventional academics might well retort "So what? We have been examining and solving problems for years and each problem is concerned with a system, within the above definition." Naturally, they would be right *but*, because of the success attainable by piecemeal attack, they have generally been working on smaller sub-systems of a larger total system. In the past, problem solvers have become expert at optimizing sub-systems and approaching wide problems by a method which consists of linear analysis, i.e. they have broken down the total system into a set of sub-systems and examined these separately without being primarily concerned about the way the sub-systems interact with each other. The new generation of systems analysts would have us believe that summation of answers about the behaviour of the component sub-systems does not necessarily tell us what the behaviour of the whole ensemble will be.

The "hard" sciences have built the major part of their considerable success on linear analysis and there is no positive reason why, when "hard" and "soft" science are compounded into technology assessment, similar success will not be achieved. Because "hard" science gave birth to the new systems analysis (out of electronics technology) an apposite analogy is possible. Electronic sub-systems were originally designed by a linear approach but very often the merging of two sub-systems introduced various deleterious multiple "feedback effects" (with the "feedback" channel formed from parts of *each* sub-system). The systems were non-linear, insofar as it is difficult to isolate the third sub-system of the feedback loop. However, when the feedback loop was detected (by checking that the sub-systems functioned correctly in separation) the whole system could be linearly analysed providing it was realized that the "new" sub-system was comprised of parts of the two original sub-systems. Nearly all systems containing interactions between ecological, economic, social, and technological sub-systems will perforce, at first sight, look non-linear. Nevertheless, it is my old-fashioned belief that unless techno-social systems eventually reveal their "linearity" under intensive scrutiny, they will remain virtually intractable to analysis of any acceptable kind. It really will not be good enough to isolate a few gross interconnections between the sub-systems which form part of a universal system

306

and then "computerize" the whole system with the aid of a series of simulated relationships which purport to represent the behaviour of the sub-systems. Such an approach is, if anything, worse than modelling the sub-systems with reasonable accuracy and then neglecting some of the less obvious interconnections between sub-systems[1].

Quite obviously, because of the way in which universities in the past have concentrated upon specialized forms of education appropriate to isolated disciplines, the advocates of change have got a very good point when they call for "systems universities" or, in Jantsch's words, "inter-, and trans-disciplinary universities". The need for new forms of university education which engender willingness to cross inter-disciplinary boundaries and solve problems concerned with interconnections between sub-systems of our society is indisputable. Those of us who have worked as analysts of technological change have suffered greatly from the early specialized training which left us festooned and cocooned in the regalia of our initial choice of discipline. Nevertheless, a needed change cannot be brought about by exclusive or excessive concentration upon the intracacies of systems analysis "theory" (as it is presently understood).

No doubt all spirited readers would like to have an attempt at designing a new university system which engenders a multi-disciplinary approach to solving societal problems, as I would myself. However, reading about the trans-disciplinary system suggested by Jantsch (1969) to the Massachusetts Institute of Technology prompted my present thoughts on this matter and, therefore, it is only fair play to reiterate some of his suggestions.

Jantsch's Trans-disciplinary University

Jantsch defines the essential characteristic of a trans-disciplinary approach as "the co-ordination of activities at *all* levels of the education/innovation system". He sees the trans-disciplinary university as comprised of three major types of unit: "systems design laboratories", "function-oriented" departments, and "discipline-

---

[1] I am not carping exclusively at Forrester and Meadows, see Meadows (1972). Naturally two concomitant and concurrent approaches, one from the general to the particular the other from the particular to the general, can be mutually enhancing. What I am carping about is the "starting of hares" based on the former without paying sufficient regard to the latter.

oriented" departments, all three to interact with each other through ubiquitous "feedback". The proposed organizational structure is based on a "relevance" system and focuses on "the interdisciplinary co-ordination of the purposeful/normative, normative/pragmatic, and pragmatic/empirical levels" of the education system. These last word-pairs quoted refer to different interconnections between levels in a "relevance tree" (see Chapter 7) which is simply an iconic diagram representing possible progress toward a normative goal. (Branches, or levels, of the tree are such that the lowest level represents empirical input, the next that of pragmatic thought, the next the normative process of connecting the pragmatic thought to the ultimate level of purposeful goals.)

In his publications, Jantsch has graphically superimposed the relevance tree upon a three-tier pyramidical diagram representing (from the appex downwards): the "systems design" laboratories, the "function-oriented' departments, and the "discipline-oriented" departments. The result is a scarabaeous diagram[1] shown below.

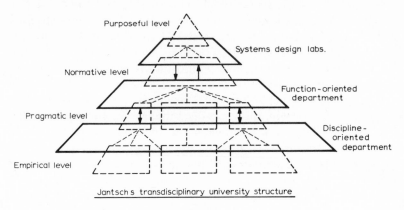

Jantsch s transdisciplinary university structure

Though their titles are fairly explicit, some further explanation of what Jantsch envisaged as the functions of his three main units of the university is necessary (even to this truncated account of his ideas).

The *systems design laboratories* were to be particularly concerned with the design of socio-technological systems including

---

[1] A "scarabaeous" was an amulet worn by Egyptians to placate a god. Jantsch's final iconic diagram may not placate many conservative university Vice-Chancellors but it has a pleasing shape and makes good sense.

the law, "hard" and "soft" science, and management. Sectorial problems were to be the main tasks allocated to the design unit, for example "ecological systems in man-made environments", "ecological systems in natural environment", "public health systems", etc. In addition to designing specific systems the design unit would be responsible for all aspects of long-range forecasting of eventualities and technologies which could affect future systems. Some limited experimental building of physical small-scale models of small systems was to be allowed (although it is difficult to perceive how this can be conveniently done when the whole purpose of systems analysis is to look at the entirety).

Function-orientated department was to take an "outcome-oriented" look at the output of technology performing in society and deal "flexibly with a variance of specific technologies which might all contribute to the same function". In the work of this department, a lot would depend upon the imagination shown in defining what society required because this invariably determines the possible technological solutions. For example, in looking at "urban transportation", the output is to move people and goods but the development of technological options depends on a clear definition of the reasons why people want to move around an urban neighbourhood.

Discipline-orientated department would comprise the conglomerate of what mainly comprises a university today. Except that Jantsch postulates that a method of building in "inter-disciplinary potential (or 'valency') of the disciplines" would be needed. No doubt the overburden of the above two units (where one would expect much of the post-graduate work to take place) would exert sufficient pressure to ensure that an eye was always kept, in the discipline-orientated department, on the opportunity for inter-disciplinary collaboration. As Jantsch points out, a worker in, say, the systems laboratory who cannot complete his task but has clearly perceived what further knowledge he needs, could have a decisive impact upon the discipline-oriented department when he returns to recharge his intellectual batteries and acquire the necessary knowledge.

As an intellectual abstraction, which one is in no immediate danger of being asked to help implement, the "trans-disciplinary" university is intuitively appealing. However, the mere presence of the systems design laboratories and function-orientated department within the customary apolitical university raises awkward problems.

It is usually tolerable (forgetting the reaction of the "mature" to youth in general) when universities presently make political comment. Nonetheless, imagine the likely reaction if a trans-disciplinary university commented adversely with apparent scientific "systems proof" of validity, on antiquidated social systems! Political control of the two upper "shells" of Jantsch's scarabaeous would then become a proper ambition for any aspiring political leader and overt attempts at gaining such control would exacerbate present day student unrest. Naturally, if the new wave of students make Jantsch's proposal an ultimate goal for the activist development of social consciousness, the universities really will become intellectual battlefields. Conversely, if government recognizes potential perils and attempts (with some justification) to control the two potentially more disruptive upper echelons of trans-disciplinary universities, the lower echelon (discipline-orientated department) can rightly claim undue political interference and, more than likely, will detach itself to become a conventional university once more.

Since the universities are, in general, funded by governments, I do not foresee Jantsch's complete proposal becoming a reality. What I do see, though, is the two upper "shells" of his scarabaeous becoming part of the structure of the executive agency of government, or part of a structure reporting directly to Parliament (as in the conceptual Office of Technology Assessment). With, either arrangement, however, I see no possibility of forging the linkages between the discipline-orientated academic system and the other two levels of Jantsch's scarabaeous. The interaction is presently as strong as it can be, in my opinion, without the possibility of creating irreversible damage to academic freedom.

More pragmatically, the expectation could be that academics will orientate some of their teaching activity towards the new needs of professionals engaged in systems analysis and technology assessment. Precursor signs that this is happening already exist and a discussion of some of these indications is the subject of the next section of this chapter.

University Teaching of Technology Assessment

There are a few universities providing courses on technology assessment, or closely related subjects, at the present time. A British pre-cursor of the university departments involved in analysis of social-technological interactions is the University of Manchester's Department of Liberal Studies in Science. This department was set up in 1966 to "devote itself to themes of the 'science-and-society' type: an aim the more unusual in that it was to be done in the context not of teaching to students in other science departments but of an Honours School based on the Department" (see Langrish *et al.*, 1972). Working within a conventional university structure, this Department has been able to pursue, at teaching and research levels, topics which are inter-disciplinary in that they cross boundaries between science, social science, and the Business School "departments" of the University. One noticeable consequence has been that a few members of staff and post-graduate students have been completely pulled over inter-disciplinary boundaries after a fruitful period in the Department of Liberal Studies in Science. If the hard-headedness of Business Schools is any criterion, this kind of absorption is not simply based on the desire of other departments to be seen to be paying homage to the interdisciplinary principle but on real need. It has also been suggested that the Department of Liberal Studies in Science has achieved a record not worse than any other department in the number of its graduates offered reasonable employment (remarks by Dr. Harry Rothman made during discussions at the OECD seminar on technology assessment, Paris, January 1972), a point worthy of note when the traditional science departments have recently had difficulties in placing their graduates in employment appropriate to their disciplines [1].

Because of their proximity to a political arena where technology assessment is considered a worthy topic, American universities have energetically undertaken technology assessment activities which may be directly related to the study of the social-technological interaction. Professor D.T. Goldman, U.S. National Bureau of Standards, initiated a course on technology assessment at the

---

[1] "Most scientists agree that if the universities provide a number far in excess of those who can fill positions requiring their specialist capabilities, there will be frustration among scientists which will dissuade the next generation from taking up a scientific career. This has long been emphasized by leading scientific figures. Part of the solution lies in arranging, as we are doing in the government service, for scientists to have far wider opportunities than scientific service alone offers and this practice, I hope industry will also adopt."

The Prime Minister, Edward Heath.
(Dalyell (1972), *New Scientist*, 54, No. 796, 18th May)

University of Maryland in 1970. Using reports by the National Academies of Engineering and Science and information provided by the Congressional Research Service, Goldman introduces graduate students to the difficulties of making objective technology assessment. Students are then encouraged to attempt their own assessments and studies already made by them include the problems of disposal of explosives in the U.S. Navy, proposal for a new port at McKeesport, Maine, replacements for the internal combustion engine, and the problems of unloading supertankers. According to the Futurist (1971), Goldman has had such success at the graduate level of teaching that it might be interesting cautiously to try teaching technology assessment to undergraduates.

Perhaps the most ambitious American university programme, related to technology assessment, has been conducted by the Harvard University Programme on Technology and Society. This programme was established in 1964 by a grant from International Business Machines Corporation to "undertake an inquiry in depth into the effects of technological change on the economy, on public policies, and on the character of the society, as well as into the reciprocal effects of social progress on the nature, dimension, and directions of scientific and technological developments".

Nearly fifty projects were initiated by the programme, up to 1970, and these projects were truly inter-disciplinary in that they were worked on by 100 representatives of 21 disciplines from 22 universities and seven industrial and public institutions. Publications emerging from the programme with its large number of personnel are, not surprisingly, myriad: about 120 articles, 30 to 40 reports and dissertations, and about 20 books. By writing to Harvard University Program on Technology and Society, 61 Kirkland Street, Cambridge, Mass. 02138 the reader can obtain cheap copies of the Programme's annual reports which are undoubtedly of good value since they contain lists of research personnel, work in hand, and publications made.

The University of California at Los Angeles (UCLA) has another imaginative approach, with Dean Chauncey Starr directing technology assessment studies within the engineering department. Starr sees technology assessment as "a systems analysis approach to providing a whole conceptual framework, complete with scope and time, for decisions about the appropriate utilization of technology for social purposes". Since Starr's participation in the National Academy of Engineering study of technology assessment helped produce many declarations of belief in the importance of

312

social values to technology assessment, we can confidently expect a new kind of "societal engineer" to emerge from the UCLA Department of Engineering.

Similar or related efforts are taking place in many other American universities and an attempt to list them all would be invidious[1]. A cursory survey of European university efforts in this field does not reveal the same concentration of effort on the re-structuring of conventional departments to fit present-day needs. Admittedly some European university departments have, in the long-standing tradition, thrown up some remarkably good individual performers in this "no man's land" between physical and social sciences. But as yet there are not many signs that there has been much conscious re-structuring of university departments to allow a cohesive departmental attack on these problems. One can expect the activities of the German Studiengruppe für Systemforschung (an organization devoted to systematic study of problems which have some area of interaction between legislation, society, and technology) to have an encouraging effect on the neighbouring University of Heidelberg and, indeed, it is rumoured that the University of Hanover is contemplating setting-up a "Sonderforschungsbereich" (research programme for "planning and organization of socio-technological systems"). Similarly we might expect the Royal Swedish Academy's interest in engineering science and technology assessment to have a catalytic effect on Swedish universities, and so on. Yet the fact remains, once again without the invidious list, that European influences forcing a change in the structure of university departments do not seem to be inspired from within the universities.

The classical claims for parallelling research with teaching activities in the universities are nowhere more justified than in this area of interaction between science, technology, and society. Whatever extremity of political stance the student adopts, he can hardly refuse to associate himself with a programme which objectively attempts to look at these interactions. He can, of course, refuse to concur with majority conclusions, but the actual process of investigation cannot fail to be appealing to all shades of political conviction. Despite the previously expressed pessimism about the

---

[1] Examples that have randomly come to my notice are

Program for Policy Studies in Science and Technology, George Washington University,

Centre for Research on Utilization of Scientific Knowledge, University of Michigan, and

Centre for the Study of Science in Human Affairs, Columbia University.

chances for the systems laboratories, and function-orientated departments recommended by Jantsch, I feel very optimistic about the chances of success for departments which teach, and research, their chosen disciplines against a background of concern about social consequences.

Future Action

Professor Arthur Bronwell, after a long and successful academic career (including being Dean of Engineering at the University of Connecticut), has created many friends and a few critics by launching upon a sea of rhetoric in his mature years. His rhetorical outbursts are aimed at the natural inertia in the institutions of society and a particularly apposite utterance is as follows.

> "The institutional structures of our society, including our universities, tend to be inordinately impacted in short-range visions and goals — the massive machinery of progress in self-perpetuating, well-structured domains of knowledge, the creative challenges of urgency and immediacy, as well as the pursuit of knowledge all the way down to the 'monumental effort on goals of trivia' — which abounds far more prolifically in our universities than we are willing to admit. Universities, learned societies, and professions have a propensity for becoming caught up in this myopic implosion in our exploding world. Our ideas and our visions tend to remain stationary and philosophically myopic, immutably constrained by preconceptions that no longer encompass more than a thin veneer of the rapidly evolving world of today — much less that of the future. In our crisis-oriented societies, human values are buffeted around aimlessly and flounder on the rocky shoals of storm-ridden seas" (see Bronwell (1972)).

Universities must certainly avoid becoming part of "the myopic implosion" yet it is not really clear how they should acquire long-sightedness and expand their activities. The trans-disciplinary university is a good normative goal but its serious implementation awaits the outcome of the present environmental—ecological debates. If the innovatory American legislation stemming from the National Environmental Policy Act (NEPA, see Chapter 2) is sustained and similar procedures are instituted in Europe, public involvement in socio-technological benefit—cost analysis will produce a more favourable political climate for the acceptance of the postulated "systems university". Nonetheless, even allowing for the probability that politicians will continue to promote the environmental lobby, it could take one or two decades before an optimally sized European "systems university" is chartered and

314

even then it will probably be formed as an international venture to facilitate the recruitment of effective members of faculty and share the risk that such an experimental venture entails[1].

Meanwhile, university faculties and students might consider nurturing the "eco-debate" with technology assessment. Such action would put some substance into the verbalization which currently passes for serious concern about environmental problems. It has been said, for instance, that the American congressional debate which led to an abandonment of the Supersonic Transport programme (SST) was conducted without either side of the political argument being in possession of an analytic briefing (see Congressional Record, 1972). Senator Edward Kennedy's initiative in stimulating

> "a group of scientists and scholars to come together and produce a book on the anti-ballistic missile which would inform the public on the issue and provide Congress with another source of expertise, with which to evaluate the Administration's proposals"

provides an example of the mechanism the university departments might adopt (see Kennedy, 1972). Naturally, if an Office of Technology Assessment existed to give direct advice to parliaments, the procedure for the universities would be simply to dissect the analyses made public. Without the stepping-stone that an OTA could provide, universities can still make and publish assessments. The only real question outstanding, forgetting the unavoidable anguish of living through a learning period when rather immature assessments will be made, is "whether research funds will be obtainable if published analyses indicate that present 'official' policy is not desirable?". Individual criticism does not have the same disruptive effect as comment which could arise out of assessment carried out by University departments. Too heavy an alignment of collective comment, supported by objective analysis, might bring down censure which could make adventurous departments yearn for a return to the ivory tower. Traditionally, the safest procedure is to carry out research for agencies of government, then there is the certain knowledge that, though these agencies will eventually adopt an "official" position with regard to any problem, the "objective" input to the "in-house" debate has been made. Some

---

[1] After thirteen years discussion, the next of the convention setting up the first University of Europe in Florence, has only recently (May 1972) been agreed. The European University will initially teach political and social science, economics, law, and history.

influence is certainly possible. Yet the debate, when the political decision is to be voted on, is at a later stage and this is the crux which the perceptive university department can anticipate. Technology assessments can be prepared to be released at the crucial time.

The steps that I would recommend university departments to adopt are as follows.

(*1*) Decide whether one wishes to exert influence by

(*a*) providing analyses to agencies of government, or

(*b*) providing analyses at the time of parliamentary debate.

(*2*) If decision (*1a*) is made, no further steps are ethically possible because the provision of advice is subject to contractual constraint. If decision (*1b*) is made, the outstanding problems are

(*a*) how to obtain information on analysis carried out for the agencies of government?, and

(*b*) how to select the areas for analysis ahead of parliamentary debate.

(*3*) Governments are prone to air their concern and the answer to (*2a*) is not difficult. In addition, a continuing perusal of the press and published "learned" papers will usually reveal areas of current concern to government agencies. Liason with departments who have made decision (*1a*) is also very fruitful. (An example of what can be found by scanning the technical literature is the information imparted, by Raymond Bowers in *Scientific American*, 215 (August 1966), on new and potentially very cheap sources of low-power microwave, solid-state, devices [1]. Carry out background technological forecasts to select areas of potential concern and institute an ethical intelligence system.

(*4*) Consider invoking government concern if an area of appre-

---

[1] The significant point about the new devices is that they will provide small microwave oscillators which are as cheap as a few dollars, as compared with current prices for bulky electron tube oscillators costing thousands of dollars. Questions arising out of proliferation of personal microwave systems (likely to be much the same as those arising from the advent of transistorized telecommunication systems) are

what is the proper use of that part of the electromagnetic spectrum that these new devices will occupy?

do these devices augur further electronic invasion of privacy?

are there potential adverse effects on health? (current standards for acceptable levels of exposure to microwave radiation differ in the U.S.S.R. and the U.S.A.: the Americans have a standard of ten milliwatts for average long-term exposure; the Russians have a standard of ten microwatts per square centimetre per day).

316

ciable benefit or disbenefit is predictable from technological forecasts. If concern is created, before parliamentary debate commences, it is difficult to avoid step (1a) but at least if this step is taken the department will be starting its analysis without the "fog" of an established agency stance. Greater service to the community would, of course, be done if a large department could trigger an analysis (to be sponsored by a government agency) whilst proceeding, at the same time, with a parallel separate analysis without agency ties.

(5) Whether University departments consider the above advice to be prudent or not, they should energetically attempt to keep surveillance over all likely impacts of system analysis on their own academic disciplines. There are only temporary advantages to be gained by arguing in general terms against the intrusion of systems analysis. Widespread adoption of systems analysis is only a question of time and its impact will be as considerable as the impact of the digital computer has been on many disciplines. One consequence of not keeping abreast of the subject is that if, due to outside pressures, systems analysis is hurriedly adopted, the result can be akin to that obtained when some departments adopted the digital computer without first developing the discrimination that the use of such a powerful tool warrants.

References

Allport, F.N. (1955). *Theories of Perception and the Concept of Structure*, New York: Wiley.

Bronwell, A. (1972). "Technological Forecasting in the Formation of National Policy", *Industrial Marketing Management*, 1, 431.

Cade, J. (1970). "The International Institute for the Management of Technology", *OECD Observer*.

Congressional Record (1972). "Establishing the Office of Technology Assessment and Amending the National Science Foundation Act of 1950", H 865 to H 886, U.S. Govt. Printing Office, Washington.

Futurist (1971). "Teaching Technology Assessment", *The Futurist*, December, 230.

Hill, K.M. (1972). "A New Approach to the Management of Technology", *Industrial Marketing Management*, 2, No. 1.

Jantsch, E. (1969). *Technological Forecasting in Perspective*, Paris: OECD.

Jantsch, E. (1972). *Technological Planning and Social Futures*, London: Cassell/Associated Business Programmes.

Kennedy, Sen. E.M. (1972). Statement made in Hearings before the Subcommittee on Computer Services of the Committee on Rules and Administration, U.S. Senate, in: *Office of Technology Assessment for Congress* published by the U.S. Govt. Printing Office to record the second session on S.2302 & H.R. 10243 held on 2nd March, 1972, p. 35.

Langrish, J. *et al.* (1972). *Wealth from Knowledge — A Study of Innovation in Industry*, London: MacMillan Press.

Meadows, D.L. *et al.* (1972). *The Limits to Growth*, A Report of the Predicament of Mankind from the Club of Rome's sponsored project, Washington D.C.: Potomac Associates.

CHAPTER 11

Future Public Involvement in Technology Assessment

"Why, in a socialist country, whose constitution explicitly says the public interest may not be ignored with impunity, are industry executives permitted to break the laws protecting nature?"
Mikolai Popov, *Soviet Life*

Introduction

Goldman (1970) argues a convincing thesis that there is a convergence of environmental disruption from both extremes of the political spectrum. Whilst there is no virtue in preparing a comparative table of the worst environmental excesses of the U.S.A. and the U.S.S.R., Goldman says is quite clear that both countries have "rivers that blaze with fire, smog that suffocates cities, streams that vomit dead fish, oil slicks that blacken seacoasts, prized beaches that vanish in the waves, and lakes that evaporate and die a slow smelly death". Activists intent upon mitigating second, and higher, order disruptive technological impacts upon economic, social, and ecological sub-environments might, therefore, pause before advocating solutions which rely upon known shifts in political wavelength. Captains of industry and government executives of all existing political complexions generally appear to be cast in the same mould.

319

Analysts in the U.S.S.R. calculate "èkonomičeskij potencial", and their Western counterparts calculate "potential economic return", of projects leading to new products and processes[1]. But in the past the public has not had sufficient access to the details of either set of calculations despite having ultimately to pay the social costs of environmental mismanagement. Neither have the Soviet analysts had any more success than Western quantifiers in attaching measurable values to social benefits and disbenefits[2]. Concerned environmentalists might consequently be forgiven if they adopt the following slogan.

"Avoid environmental stress, just strain technology's *covert* assessors down the post-Malthusian drain."

This chapter contains suggestions as to how the European public might organize themselves to use assessment techniques in disputations about decisions which might affect their future environment.

Portents of Future Development in the Style of Public Discourse

There would not be much need to advise environmentalists to brush up on their knowledge of the techniques of technology assessment if it was thought that the quality of "official" assessments would remain at the standard publicly exhibited in the United Kingdom during debate on the siting of London's third airport. In order to draw out analysts who operate at such a facile level, it appears quite adequate to conduct vigorous verbal arguments against suspect quantification[3]. At the same time, it is not to be expected that such easy victories as that won over the Roskill (1970) Commission Report are going to be won in debates about rather more subtle aspects of environmental concern. (One needs limnologists, for example at Lakes Baikal and Erie, to define

---

[1] See Zaleski *et al.* (1969) for an account of some methodological practices in the U.S.S.R.

[2] Rothman (1972), in his study of pollution in industrialized society argues, as a Marxist, that Soviet-style bureaucracy and capitalism are the main obstacles to solving environmental problems. But not even Marxists will be able to avoid quantitive assessment and balancing of economic social and environmental benefits and disbenefits.

[3] See the celebrated satire by John Gordon Underwood Adams (1970), and Adams (1972) on the schizophrenia inherent in some kinds of quantification.

320

degrees of pond and lake pollution.) And assessors working for large companies and executive agencies of government are likely to become much more sophisticated in their presentation of environmental analyses *after* there has been public outcry against certain changes.

Oliver Goldsmith gave us some indication of what to expect when he wrote in Letter CXI of the *Citizen of the World*

> "Hunters generally know the most vulnerable part of the beasts they pursue, by the care which every animal takes to defend the side which is weakest: on what side the enthusiast is most vulnerable may be known by the care which he takes in the beginning to work his disciplines into gravity, and guard them against the power of ridicule."

Environmentalists must expect "official" analysts to work their disciplines into gravity by use of some seemingly powerful methodology against which redress will not always be gained by purely verbal argument. My own scenario for the future is based upon the hypothesis (used in Chapter 1) that increments in the degree of perceived risk, threatened by external forces, give rise to enhanced activity by the assessors and innovators of technological change. All protagonists of technical change are now aware of the threatened risk to some (not all) technological enterprises which is posed by the environmentalists' lobby and will take action accordingly. Some advocates of technological plans will be influenced beneficially but there will be others who will naturally be tempted to circumvent legitimate objections to their plans.

European environmentalists have excellent precursors before them in the form of the NEPA Section 102 Environmental Impact Statements (see Chapter 2) and a perusal of these should convince them that European debates are likely to be, after an initial period of amateurish confusion, very much more articulate, scientific, and numerate[1]. The imminent (at the time of writing) UN Conference on Human Environment, to be held in Stockholm, has also prompted European agencies of government to think more carefully about our environment and Tinker (1972) claims that Peter Walker's Department of the Environment is indulging in thought which "can threaten the mutual interdependence of ministeries". Reorientation of this kind within Government should produce

---

[1] The early impact statements are not perfect models of what a technology assessment should ideally be. Of the hundreds submitted to the Council for Environmental Quality, some are puerile, others exhibit quality, but the above-mentioned trend exists.

better analyses of environmental problems with, of course, the usual imitative but less reliable offerings.

British industrialists have also been quick to note current trends and have decided to participate in a "joint CBI—government working party which is already preparing a voluntary scheme whereby a committee will vet chemicals, and try to evaluate and thereby guard against adverse environmental effects". Naturally all serious observers will applaud this move because the Confederation of British Industries is entering upon this joint venture in the public interest. Yet, remembering the adage "if you can't beat 'em, join 'em, and if you can't join 'em imitate them", we should include in this scenario a slight probability that a government department and a national association of industrialists could be in cahoots in promulgating a detailed, but erroneous, analysis of a new technology: what, then, would be the chances of environmentalists effectively making their point of view without recourse to their own detailed technology assessment?

There is no bogy of machiavelian European governments in partnership with wicked industrialists. In fact both are reacting constructively at present but as we have seen, for example with pharmaceuticals, terrible havoc can be caused by the well intentioned. Now that the environmentalists' lobby has produced growing political and industrial action, it is time to consider the next phase when the environmentalists will have to put up their own contributory assessments or shut up. Environmentalists who stand up in the future and want to be counted must make themselves visible by donning analyst's garb. The days of the all singing, all dancing, verbalizing environmentalist are nearly over!

Categorization of Types of Environmentalists

As with any other cross-section of the public, the people promoting environmental vigilance from outside the establishment will fall into many categories. The U.S. Committee of Government Operations[1] neatly categorized them as "idealists", "realists", "technocrats", and "legalists".

---

[1] The Environmental Decade (Action Proposals for the 1970's), Twenty-Fourth Report by the Committee on Government Operations, 91st Congress, 2nd Session, *House Report No. 91—1082*, U.S. Government Printing Office, Washington D.C. 20402. (Price 20 cents.)

The suggestion is that "idealists" contend that only when individuals fully comprehend "their personal dependence on, and obligations to, the environment will the ultimate solution to degradation and abuse be found". Acceptance of such a statement inevitably leads to greater emphasis being placed upon the responsibility of educationalists, who have the power to teach a less laisser faire approach. And not only does this educational approach have to affect schoolchildren, it can, and does, intrude into adult education. Shareholders of Rio Tinto Zinc at their annual general meeting in 1972 were supplied, by the "Counter Information Service", with tracts which attempted to influence their attitude towards proposals for open-cast mining of copper in British national parks and shortly afterwards the B.B.C. transmitted a television programme in which some of the pros and cons of Rio Tinto's possible scheme were aired. Rio Tinto, with commendable acumen, have riposted with the statement that their future plans depend critically upon the findings of an independently commissioned study of the possible impacts of their potential mining programme.

"Realists" are said to contend that "individuals and industries will seldom follow sound principles of environmental management on moral grounds alone". Realists tend to throw up practical suggestions like "Why not impose a tax on pollution, whose severity is graduated over time to allow industry to make the desired transition without unduly harming the economy?".

"Technocrats" are bracketed as arguing that what science and technology have caused by way of environmental degradation, science and technology can clean up. In general, technocrats favour the technology assessment approach which is based on the underlying premise that more science and technology need not be harmful (and can be positively beneficial) if technocrats apply their craft with a sense of social responsibility.

"Legalists" are defined as those who have developed the concept that "formal protests through lawsuits to prevent detrimental environmental encroachments, and more aggressive enforcement of existing laws including the imposition of heavy fines by the Government for failure to meet established control standards, are the most effective tools for preventing environmental decline".

My own opinion is that, whatever the differences of approach, all the above categories of environmentalists will limp away from future disputes unless they enter the debate supported by the intellectual crutch of a properly prepared technology assessment.

323

Legalists cannot expect the judiciary to make decisions on environmental alternatives without knowing what possible outcomes have been assessed. Lord Hailsham, the Lord Chancellor, in the 1972 Holdsworth Lecture has surprised his critics by appearing to be in full agreement with them when they say that judges may be getting out of touch with technical matters and that it would not be a bad idea if they were assisted by experts of one sort or another (see New Scientist, 1972).

Idealists will find that any concerted educational programme that does other than induce a needless phobia against all change, will need to emphasize skill in the evaluation of the outcomes of perturbations in the delicate balance between all three (economic, social, and ecological) parts of the environment. Inexorably this will lead to the need to teach carefully graduated aspects of technology assessment technique. Yet opposition to this need already exists: Mr. Maurice Ash, chairman of the Town and Country Planning Association, has already said that environmental education is "dynamite" and that critics fear the subject is concerned neither with teaching or learning[1]! (See The Guardian's 2nd June, 1972, report on the International Federation of Housing and Planning world congress.) However, Mr. Ash has the right idea when he says "If it is to be anything at all, it must be concerned with doing. Children, therefore, must partake in the controversies, whether about polluted streams, the impact of motorways, the siting of reservoirs. They must do surveys and evaluate the results, that is if we want the environment to enter into their education. What this might do to the schools is a different question."

The realists will also need to be complemented with technology assessments of what it is they are being "realistic" about; they cannot, in good conscience, advocate that certain kinds of force majeure be applied to polluters without first defining types, degrees, and consequences of pollution. They must also understand

---

[1] The fear seems to be that Environmental education is ground to challenge the establishment and "in the meantime the establishment was trying to take it over and tame the issues". Nevertheless the Declaration on the Human Environment, submitted to the U.N. (1972) Stockholm conference recommends

"Education in environmental matters, for the younger generation as well as adults, giving due consideration to the underprivileged, is essential to broaden the basis for enlightened opinion and responsible conduct.... in protecting and improving the environment....".

324

what the effect of imposed constraints will be on all three parts of the environment.

Suggestions for a Phased Programme of Environmentalist Action

Given that there is some substance in the above comment environmentalists will need a forward-looking plan. A sufficient objective for this plan could be:

Maintain not less than parity in technology assessment technique so that disputations between public and sectorial interests are not unbalanced[1].

The big battalions of industry and government agencies are not easy to match in the provision of assessment expertise because they have the full-time staff with supporting facilities and money to spend. By comparison, an activist lobby has only got the potential that size and wide coverage of the available population can bring. Nevertheless, if it is recognized that there is likely to be equality of basic intellect on both sides, the numerically larger lobby can organize to good advantage. (With similar initial disadvantages, organized labour has managed, some would say too successfully, effectively to counterbalance management.)

One would expect the executives of an environmentalists' lobby to judge the optimal size of their association by estimating how many members, with only part-time effort, are needed in each discipline to match the analysts and advocates representing vested interests of industry or government executive agencies. Despite the greater uncertainties and lack of cohesiveness of voluntary associations, part-time contributors can achieve more than parity (in, say, economics or detailed knowledge of procedures for the disposal of toxic materials) by purely equating available working time. If the average volunteer says he thinks he can contribute a quarter of his time then four volunteers equal slightly more than one full-time expert. Allowing for the maxim that "two heads are better than one, even if one is a cabbage" some relaxation of direct parity in the cumulative working time may be made, even if it is acknowledged that the part-timers have less time to generate and communicate a fruitful idea. By equating the number of useful interac-

---

[1] There is a perceptive comment on this point by Bernard Dixon in the New Scientist of 25th May, 1972. Good reportage on environmental problems is usually found in all copies of New Scientist.

tions between analysts, it is possible to derive a very crude rule-of-thumb[1].

Number of analysts required
by the environmentalists $= \dfrac{[4x^2 n(n-1)-1]^{1/2}-1}{2}$ $(n \neq 1)$
("environuts" or "econuts")

where n is the number of full-time analysts one wishes to oppose and each part-time analyst has only $(1/x)$th of his time available to give service to the environmentalists' lobby.

Examples of the advantage of larger numbers are

(a) *When n = 4 and (1/x)th is one half* then the above formula says approximately seven econuts are required as opposed to eight on direct parity; however, note the advantage when the numbers increase.

(b) *When n = 4 and (1/x)th is one eighth* then the above formula says approximately twenty eight econuts are required as opposed to thirty two on direct parity.

The above formula is only used to point to possible tactics and should never be used if $n = 1$ (i.e. only one "official" assessor is to be opposed).

The magnitude of the numbers of specialists required in the opposing lobby is going to be very large indeed when one realizes how many full-time experts the vested interests can draw upon. Viable associations of econuts in conflict with, say, the U.K. Department of Trade and Industry *and* the Department of the Environment would need to be as large as a minor political party. Which is one reason for thinking that the present concern for the environment may lead ultimately to a reorientation of politics. Unfortunately, many of the specialists an "ecolobby" would wish to co-opt are already in the employment of some of the vested interests and in the above rather fanciful example (who would really believe that both DTI and DoE would be working against the public interest?), the workings of the Official Secrets Act would mean that government employees could not lend their support. Nonetheless the suggestion to "environuts" is that they

(1) Form larger associations.

(2) Categorize their membership by expertise (this does not

---

[1] This formula is based on the likelihood that the full-timers will have not less than $\frac{1}{2}n(n-1)$ useful interactions. For the part-timers, the formula is divided by x^2.

mean an ecolobby is the province of the professional classes. Artisans had more knowledge of the malpractice in cyanide-dumping in the U.K. than any other sector of the community).

(*3*) Employ full time secretariats to co-ordinate communication between volunteers and select appropriate representatives to fight particular causes.

(*4*) If possible, raise funds by subscription which are sufficient to maintain adequate surveillance over all three environments and publish a newsheet or journal for the membership.

(*5*) Campaign for evening courses to be run on the techniques of technology assessment and campaign, through parent—teacher associations, for elementary courses and field work on technology assessment in the schools.

(*6*) Attend political meetings to test the real strength of the political candidate's feelings about environmental problems.

(*7*) Do not automatically assume that the requirements of the ecological environment are predominant. Acknowledge that, on occasion, the other environments (social and economic) can take precedence.

(*8*) Remember that the moral justification for an envirolobby is to help in the "renewal of our basic decision making institutions" and not the protection of a vested interest against the requirements necessary for the realization of overall communal benefit.

(*9*) Beware of take-over by existing political creeds and parties.

Conclusion

The above diatribe has been made notwithstanding my belief that the "official" institutions of government are the best hope for the avoidance of any possible environmental calamity. It should not be inferred that I distrust the official system; I have simply been playing "devil's advocate" in what I hope is a constructive way.

References

Adams, J.G.U., (1970). "Westminster: The Fourth London Airport?", *Area*, No. 2.
Adams, J.G.U., (1972). "You're Never Alone With Schizophrenia", *Industrial Marketing Management*, 1, No. 4.

Goldman, M.I., (1970). "The Convergence of Environmental Disruption", in: *A Reader in International Environmental Science*, Environmental Policy Division, Congressional Research Service, Library of Congress, 60—116, U.S. Government Printing Office, Washington D.C.

New Scientist (1972). "Feedback: Advising the Judges", *New Scientist*, 54, 1st. June, 516.

Roskill (1970). *Report of the Roskill Commission*, London: Her Majesty's Stationery Office.

Rothman, H. (1972). *Murderous Providence*, London: Hart-Davis.

Tinker, J. (1972). "Nemesis and the Constitution", *New Scientist*, 54, 25th May, 444.

Zaleski, E. *et al.* (1969). *Science Policy in the U.S.S.R.*, Paris: OECD.

CHAPTER 12

Future Involvement of Industry
in Technology Assessment

"The first objection that technology assessment can't pay for itself, is, for some industries, as short-sighted as maintaining that market research per se produces no profit."

Kirchner and Laserson (1972)

Introduction

Some industrial companies in Europe and America have suffered from what has been described as "environmental intransigence" and will (as remarked earlier) make the necessary, ameliorating, technology assessments in the future. More fortunate companies who have not yet been affected by the environmentalists' lobby, believe that the whole question of technological impacts is solely an American concern which will not impinge upon their business interests in Europe. Companies already trading in the U.S.A. take the view that only their American business will be threatened by the U.S. legislation to set up an Office of Technology Assessment and tend to rely upon a "legalist" defence which may be unnecessary but is best summarized by the following statement (taken from my correspondence and quoted anonymously for obvious reasons).

"The arrogated right of an Office of Technology Assessment to serve

329

subpenas to aliens for attendance to give testimony as well as for the production of documents located abroad is tantamount to a serious impingement on the sovereignity of the nation state to which the person subpenaed is a national." (comment made by a company who examined Bill HR 10243 before the Brooks amendment altered clause 6, see Chapter 2)

"Therefore such conduct constitutes a breach of international public law against which the governments of the nation states concerned should protest vehemently. In point of fact, various governments have already enacted defensive legislation designed to prohibit their nationals from obeying injunctive orders from foreign authorities. However, it is doubtful whether these various legislations would also cover subpena orders as provided for in the (unamended) new American Bill HR 10243. Since it can hardly be anticipated that this Bill will be stopped in the U.S. Senate, the governments concerned should be made aware of this situation in order to enable them to extend their present legislation or to enact adequate legislation in order to protect their nationals against such subjection to the power of the U.S. administrative authorities."

Very few companies take the longer view that, as we chronologically proceed towards the "post industrial" society, Business could become "one of the central 'knowledge institutions' which some sociologists expect to become the predominant new type of institution" (see Jantsch, 1972). Slightly more companies (but still a few) take the entrepreneurial view that technology assessment (by its very nature an expensive exercise) presents opportunities for profitable contract business with government.

This chapter leaves aside discussion of industrially produced assessments aimed at avoiding too much "environmental intransigence" because it is fairly obvious that such strategy depends almost exclusively upon being forewarned (a new function for industrial intelligence?) and emphasis upon quantification of the economic and social benefits which are often put into the scales to counterbalance environmental disbenefit. The concentration in this chapter is on summarizing the rather more positive approach to technology assessment that might be possible if industry is encouraged to sell its assessment services directly to government.

Assessment Contracts for Industry

In the U.K. it seems as if the public service has an advantage over the private sector in the practice of technology assessment. Naturally, such a conclusion may be erroneously generated by the public servant's greater plausibility at conferences and seminars

(see Peter and Hull (1971) for a description of Peter's placebo —
"an ounce of image is worth a pound of performance"). However,
it is certainly true that larger government departments have greater
opportunity to exercise their skills on the multi-faceted problems,
simply because industrial employees are normally asked to perform their analyses against the companies' narrower, commercial,
criteria. Such a situation is as unhealthy as that where the environmentalist (as an individual) cannot match the assessment skill of
government agencies and private business.

Quite obviously one cannot expect industry to invest altruisti-
cally the large expenditure necessary to make comprehensive assessments, including social and environmental impacts, to compli-
ment their straight-forward analysis of the market potential of
technological goods. On the other hand, in the case where technology rests upon industrial innovation, the company concerned
(or a similar company with smaller vested interest) is best placed
to begin the technology assessment. Government assessors (like
the Programmes Analysis Unit) do, of necessity, spend a lot of
time familiarizing themselves with the technology, usually with
the relevant industry acting as mentor. Because technology assessment is often concerned predominantly with the longer term impacts, there are good arguments for utilizing industry as contact
assessor to government. By studying the longer term impacts, a
company gains only marginal advantage (mainly in analytical experience) over competitors and even this modicum of lead is greatly
reduced when the assessment is published. With governments
awarding large production and design contracts to industry, it is all
the more surprising that they have not used companies more to
provide equally valuable high-quality technology assessments. Can
the reason really be that, outside their own executive departments,
governments place more importance upon the production of tangible (if not always suitable) hardware? Why are governments willing
to spend hundreds of millions on, say, civil aviation and yet baulk
at expenditure, smaller by several orders of magnitude, on cogent
assessments? Perhaps reluctance to finance more than a minute
number of technology assessments stems from the belief that to
contract out assessment is to abrogate some power of decision
making? Yet, in fact, decision makers' powers can be enhanced by
improving the quality of assessments and analytical advice received
before taking a decision.

In the U.S.A., shortly after the President's Executive Office
showed an interest in technology assessment, the President's Of-

331

fice of Science and Technology (OST) commissioned a $100,000 study from the MITRE Corporation for an exploratory technology assessment (see Jones, 1972, and Strasser, 1972). The OST's perceptive recognition of the "methodology gap" between what appeared a priori possible (and plausible) and what had been accomplished led to the issuing of the contract[1]. Gabor Stasser of OST described the task, on which the OST/MITRE development of technology assessment methodologies was to be based, as

"seeking some way by which all of us will better understand the interrelationships and trade-offs of the many different objectives that we seek, and the many different costs that we must incur, in a language that permits dialogues among policy and decision makers, their constituencies, including the intellectual community, industry, and analysts, or in short, all those who are involved in one way or another in our collective actions as a society".

With this last objective in mind, OST/MITRE selected five areas for "pilot study" and proceeded with "the identification and development of a methodology for technology assessment".

As might be expected with an expenditure of $100,000, considerable effort went into the project, the outcome of which is outlined below. (Readers interested in obtaining "cheap" spin-off from the OST/MITRE work can obtain a fuller six-volume account with summary from the National Technical Information Service, Springfield, Virginia 22151, if they use the order number PB 202 778 and enclose $31.50 plus extra postage for delivery to Europe.) The complete OST/MITRE report includes the following sections:

Some Basic Propositions by Martin V. Jones
Automotive Emissions by Willis E. Jacobsen
Computer-Communications Networks by Hugh V. O'Neill
Enzymes (Industrial) by David R. Rubin
Mariculture (Sea Farming) by Robert C. Landis
Water Pollution (Domestic Wastes) by Victor D. Wenk

Seven "major generic" steps in making a technology assessment were identified in the OST/MITRE contract.

---

[1] The phrase "methodology gap" could evoke in European minds the suspected gap between the analytic capabilities of the government and industrial sectors but I do not expect the U.K. Cabinet Office to commission separate studies from each sector in order to compare potentials and techniques of technology assessment.

332

(*1*) *Define the assessment task*

(*a*) Discuss relevant issues and any major problems.

(*b*) Establish scope (breadth and depth) of inquiry.

(*c*) Develop project ground rules.

(*2*) *Describe relevant techniques*

(*a*) Describe major technology being assessed.

(*b*) Describe other technologies supporting the major technology.

(*c*) Describe technologies competitive to the major and supporting technologies.

(*3*) *Develop state-of-society assumptions*

(*a*) Identify and describe major non-technological factors influencing the application of the relevant technologies.

(*4*) *Identify impact areas*

(*a*) Ascertain those societal characteristics that will be most influenced by the application of the assessed technology.

(*5*) *Make preliminary impact analysis*

(*a*) Trace and integrate the process by which the assessed technology makes its societal influence felt.

(*6*) *Identify possible action options*

(*a*) Develop and analyse the various programmes for obtaining maximum public advantage from the assessed technology.

(*7*) *Complete impact analysis*

(*a*) Analyse the degree to which each action option would alter the specific societal impacts of the assessed technology discussed in step (*5*).

Two very short summaries, taken from Strasser (1972), of the outcome of two of the five "pilot studies" are as follows.

Automotive Emission Controls

"This study analyses the automobile's role in causing air pollution from the standpoints of the kinds of pollutants emitted, the amounts emitted, and the damages inflicted as a result of these emissions. The study reviews automotive emission control technology which would help abate automotive-generated pollution, including devices which modify emissions produced by internal combustion engines, alternative power sources (e.g. electric power plants), and institutional factors which would reduce automotive-generated air pollution (e.g. driving bans within central business districts)."

"The study identifies and defines the state-of-society conditions which are directly and indirectly related to automotive emission control technology. Among the topics covered are: present and future emissions, emissions criteria, air quality standards, air pollution monitoring, automotive use and demographic trends, urban mass transit technology and patronage, and emission control costs. The time covered is 1965—1985."

"This study concludes that large societal impacts may result from seemingly small changes in technology or in emissions control actions. The study provides an analytical framework for relating the costs incurred in reducing automotive-generated air pollution with the costs saved in the resulting improvements in air quality, health, and national welfare."

Enzymes (Industrial)

"The enzyme industry is still in its infancy — scientists have crystallized about 100 pure enzymes, have described the characteristics of about 1,000, and believe that there may be as many as 100,000. Due to the relative newness of the technology, this study gives major emphasis to understanding the enzyme process. Topics covered include: the direct and indirect uses of enzymes in products, uses of enzymes secreted by microorganisms, and production and technological innovations. Forecasts of trends and developments in this technology are made to about 1980 and 1990 time periods."

"Some of the applications of enzyme technology are considered to be replacement of high temperature and pressure processes, lower-cost sugars, and new and improved pharmaceutical products. This study suggests the possibility that some of the applications could have adverse consequences. For example, an enzyme process could displace cane and beet sugar as the primary source of sweeteners could have demographic and international political consequences in terms of displacing workers and changing our protectionist policy toward foreign producers of cane and beet sugar."

"Some of these applications could have large impacts on a complex variety of phenomena, such as energy consumed, pollution, capital equipment purchased, and the spectrum of viable products. This study provides a framework for the analysis of such impacts and discusses various control options which would minimize the adverse impacts of enzyme technology."

The OST/MITRE study closes with some specific recommendations see Jones (1972), which are summarized as follows.

"Better methods should be developed for the selection of the technologies to be assessed, because only a fraction of the money and trained man-power will be available for the assessments that are needed.

For the same reason, better ways are needed to decide what potential impacts should *not* be analysed.

Forecasting methods need to be improved. Because expert opinion is usually the major information source in making futuristic assessments, there should be research into objective methods of eliciting and analysing different expert opinions. Also needed are more tractable, objective methods for generating scenarios, or descriptions of how a situation may develop. Currently it is all but impossible to decide which of two conflicting scenarios, starting from the same information base, is the more reliable.

The nation needs to expand the supply of basic data used to trace the sequence of impacts in an assessment. We need to understand *lag—lead* relationships (how soon one event will follow another) and *magnitude* relationships (how much one factor will change in response to a specified change in another factor).

The general methodology that this study developed should be adapted to the specific assessment needs of various government agencies.

Since decision making in our pluralistic society is widely distributed, ways must be found to communicate the findings of complex technology assessments to millions of citizens in terms that they can understand."

Most European governments contain departments which have in the past provided analyses which go some way towards equalling the range of the OST/MITRE study and possibly some departments have (on occasion) surpassed the quality of this experimental American technology assessment. Yet even where there are large departments, like Britain's Department of Trade and Industry, there have been no organized attempts at harnessing the resources of industrial companies to make forward-looking assessments. Since national industries are becoming larger and less fragmented, it would be relatively easy to commission ongoing technology assessments from each sector of private industry without creating too much parochial resentment. Even if the assessments tended to be reasonably biased, by the industrial ambitions of the contractors, the outcomes of the analyses would still be of significant interest to the executive arm of government. A trite example here is that no industrial sector could, without fear of drawing criticism upon itself, be unduly abusive about the consequences of lack of government foresight if the industrially produced assessment records the same lack of vision! The "look out" function of government would also be considerably improved by such commissioned technology assessments. The Government would not be foisted with placebos because a good civil servant can normally detect, and throw into the balance a counter-argument about, a biased prognosis. An exchange of assessments between government and industry can but be beneficial even if governments have partially, or wholly, to foot the industrialist's bill. Closer liaison of this sort might also do more to encourage professional civil servants to diffuse into industry. The cost of maintaining such joint assessment programme might, indeed, be less than the incremental cost of ensuring adequate governmental control over the technological impacts through the recruitment of ever increasing numbers of specialists to the public service.

The above remarks apply, of course, to the relationship between industry and the executive arm of government. But given the existence of an Office of Technology Assessment (serving the legislative arm) the system of commissionning industrial studies has even more to recommend it. If the executive agencies had a statutory duty to produce their own and commission industrially produced assessments, an OTA could effectively hold the ring with the parliamentarians acting as final judges. The major arguments against such a procedure are that it could be too expensive and that leakages of commercial secrets would ensue. When governments spend such large amounts on what the public sometimes regards as technological follies, then the first argument has little validity because the suggested expenditure on more assessments of technology could indeed lead to a saving by government and an increase in return to industrial capital. The second argument is more difficult to deny. But technology assessments are not likely to disclose many industrial secrets which could be of benefit in the immediate future and the activities of the covert industrial intelligence agencies present a much greater *immediate* security problem. Arrangements might also be possible to ensure automatic governmental protection of fairly arrived at ideas of commercial significance which are aired in commissioned assessments.

Finally, the encouragement of industry to do technology assessments for government with financial support from the public chest might make it easy to implement the useful suggestion by Tribe (1971) (see Chapter 2) that there should be a proper market adjustment with "the establishment of substantial economic rewards for the socially useful technological development, parallelling the patent system but clearly identifiable from it....". Legislation and implementation of such a suggestion could lead very quickly to industrial companies gaining honorable profits and public esteem as the knowledge institutions mentioned earlier in the introduction to this chapter.

In the absence of a properly thought out cohesive partnership in technology assessment between government and industry, it follows that small consultancies can quickly "skim the cream" off any funds which government, or industry, cares to make available as a token gesture. No one can blame consultants for such behaviour and the entrepreneurs amongst them need no advice from outsiders on this matter.

References

Jantsch, E. (1972). "Enterprise and Environment", *Industrial Marketing Management*, 1, No. 2.

Jones, M.V. (1972). "The Methodology of Technology Assessment", *The Futurist*, 6, No. 1, 19.

Kirchner, E. and Laserson, Nina A. (1972). "Technology Assessment at the Threshold. An Innovation Special Report: Technology's New Political Environment", *Innovation*, No. 27.

Peter, L.J. and Hull, R. (1971). *The Peter Principle*, p. 150, London: Pan Books Ltd.

Strasser, G. (1972). "Methodology for Technology Assessment", *OECD Seminar on Technology Assessment, 26—28th January, 1972*, Paris: OECD.

CHAPTER 13

Loose Ends

Introduction

It was not obvious during the writing how the twelve chaptered loose ends of this book should be drawn together. There was also apprehension that, through exposition of known technique, inadvertent encouragement would be given to selfish manipulators of technology assessments. Now some problems are solved, and fears calmed, by the Reuter's report on the "declaration on the Human Environment" in which delegates to the U.N. Conference on the Human Environment expressed their aspirations during their Stockholm meeting in June, 1972.

In this last chapter loose ends will be drawn together in summaries categorized by the twenty five principles of the Declaration and it is hoped that the reiteration of these principles will give encouragement to participate in, and not manipulate, technology assessment. For ease of numbering and to avoid confusion, the U.N. principles will be listed in their original order. Thus some cross referencing of the underlying comments will be necessary.

Principle 1

"Man has the fundamental right to freedom, equality, and adequate conditions of life, in an environment of a quality which permits a life of dignity and well-being, and bears a solemn responsibility to protect and improve the environment for present and future generations. In this respect, policies promoting or perpetuating apartheid, racial segregation, discrimination, colonial and other forms of oppression and foreign domination, stand condemned and must be eliminated."

Acerbic policies which promote oppression are not always as clearly manifest as those listed in principle 1. When policies have technological content, their impact upon social economic and ecological environments of present and future generations is not foreseeable without carrying out detailed technical analysis. Such analysis of possible impacts has been labelled technology assessment. It may well be that promoters of techno-policies will become, more and more, members of multinational companies without clearly defined allegiance to nation states. A State's parliament and citizens may, therefore, stand in greater need of cogent and timely information on the predictable impacts of future and present technology than they did in the past.

Discrimination and racial segregation, for example, may be deliberately or inadvertently brought about by technological means without the overt use of oppressive political policies. By structuring the use of educational technology to favour a particular sector of the population and teaching a privileged hierarchy the new dominant skills which will be required for the future, normatively planned economy, their dominance can be virtually assured. And even hierarchical prejudices can be re-inforced by technological means.

The technological tools are now available to would-be manipulators of an Orwellian world and these should be assessed with as much concern as the actions of those beyond the political pale are studied.

Principle 2

"The natural resources of the earth, including the air, water, land, flora and fauna, and especially representative samples of natural ecosystems, must be safeguarded for the benefit of present and future generations through careful planning and management, as appropriate."

The "careful planning" alluded to above needs to be very carefully organized and this is difficult because we are only beginning to understand how technology assessments should be carried out. However, it is very clear indeed that planning cannot be safely left solely in the hands of executive agencies of government and private industry. Some precautions against bureaucratic impedance and vested interests, in procurring profit from despoliation of the environment and its reserves, are necessary. To this end the best legislation known to me is contained in the National Environmental Policy Act and the Bills HR 10243 and S 2302, designed to establish an Office of Technology Assessment for the U.S. Congress (see Chapter 2). Although the legislation is naturally structured for the United States, it is sufficiently generalized to form a useful model for action by other nations, particularly the European Community and its member states.

Principle 3

"The capacity of the earth to produce vital renewable resources must be maintained and, wherever practicable, restored or improved".

Palliatives for the "depletion syndrome", or excessive fear of irreversible depletion of the earth's reserves of resources, may be obtained from more accurate assessment measured and unmeasured reserves. Lasky of the U.S. Geological Survey, has demonstrated how dependent the quantity of reserves is upon the acceptable tenor of the resource and, in particular, he has shown that a reduction of 0.5% in tenor leads to a three-fold increase in the U.S. copper reserves. (See Landsberg *et al.* (1960) on resources in America's future and Medford (1969) on marine mining in Britain.) Since tenor of a resource is a function of the extraction and processing costs, reserves of the earth's resources are a function of our technological skill up to a very large finite limit on reserves. Ultimately, therefore, if we really are space-ship earth, application of principle 3 depends upon our technological skill in using and recycling what we have. For example, the most complete solution to our recycling problem may be the apparently utopian idea of nuclear dustbins where, by converting used resources into highly ionized plasma in the centre of a fusion reactor, it is theoretically conceivable that the basic elements can be separated, recycled, and re-used. Technology can, given sufficient co-operation from mother nature, provide us with fusion reactors which could solve our energy-deficit and re-usable resource problems. Science and technology should, if the worries of the environmentalists are justifiable, be evaluated against criteria wider than the criterion of simple economic benefit—cost. Good technological

340

husbandry predicates greater effort on better, if not more, technology.

Principle 4

"Man has a special responsibility to safeguard and wisely manage the heritage of wildlife and its habitat which are now gravely imperilled by a combination of adverse factors. Nature conservation, including wildlife, must therefore receive importance in planning for economic developments."

Since wildlife is inherently part of the undisturbed environment, legislation like that of NEPA (see principle 2 and Chapter 2) provides for the necessary protection of natural species. The European enactment of similar legislation and its scrupulous enforcement is essential.

When benefits of development are claimed to be so high that destruction of wildlife and its habitat is advocated, the public should be given a democratic opportunity, through legislation like NEPA, to oppose this conclusion.

Principle 5

"The non-renewable resources of the earth must be employed in such a way as to guard against the danger of their future exhaustion and to ensure that benefits from such employment are shared by all mankind."

Principles 3 and 5 are obviously related and the comments made under principle 3 are, therefore, relevant here. Furthermore the definition of non-renewable is continually changing according to the direction in which technology progresses. The chimerical "fusion-dustbin" and less esoteric reprocessing technologies would ensure that many "non-renewable" resources would become renewable.

A first step in sharing of benefits is to ensure that the benefits are perceptible to all, including those who are expected to bear the greatest social cost of the programme leading to benefits. The establishment of Offices of Technology Assessment free of vested interests could ensure that the public became benefit and cost conscious. Perusal of the preceeding chapters might germinate dormant consciousness that things are presently not as they should be and could be considerably improved by sectorial effort in the

341

universities, industry, public service, and voluntary associations (see Chapters 3 and 10—12).

Principle 6

"The discharge of toxic or other substances, and the release of heat, in such quantities or concentrations as to exceed the capacity of the environment to render them harmless, must be halted to ensure that serious or irreversible damage is not inflicted upon ecosystems. The just struggle of the peoples of all countries against pollution should be supported."

Once again we are back to the need for protection through Environmental Policy Acts and Offices of Technology Assessment. The definition of what is toxic and what concentrations are harmful is a highly professional part of assessment procedures. But this necessary need for professionalism should not be adduced to support only paternalistic assessments by industry and the executive agencies of government. Parliament and public should have more direct information from their own assessment agencies. The setting-up of OTA's and the enactment of environmental policy acts would give clear evidence that the struggle against pollution is being supported by other than vote-catching rhetorical means.

Principle 7

"States should take all possible steps to prevent pollution of the seas by substances that are liable to create hazards to human health, to harm living resources and marine life, to damage amenities, or to interfere with other legitimate uses of the sea."

With principle 7, a need arises for international OTA's because national OTA's may reasonably take a biased view when a large national benefit can be achieved with concomitant costs and disbenefits spread internationally. Since the U.N. Conference voted for a ban on the ocean dumping of toxic wastes, we might, optimistically, hope that agreed international law on pollution of the sea will be improved. But research is needed to enable the law makers to enact laws which will cover pollution hazards stemming from new technologies which are now only dimly perceived. (Note the section on the law and technology assessment in Chapter 2, and Chapter 7 on technological forecasting.)

342

Principle 8

"Economic and social development is essential for ensuring a favourable living and working environment for man and for creating conditions on earth that are necessary for the improvement of the quality of life."

Technology is embedded in economic and social development; thus all new developments need to be technologically assessed in the way defined by Huddle, quoted in the preface to this book: "Technology assessment is the purposeful, timely, and iterative search for unanticipated secondary consequences of an innovation derived from applied science or empirical development, identifying affected parties, evaluating the social, environmental, and cultural impacts, considering feasible technological alternatives, and revealing constructive opportunities, with the intent of managing more effectively to achieve societal goals".

Principle 9

"Environmental deficiencies created by the conditions of underdevelopment and by natural disasters pose grave problems and can best be remedied by accelerating development through the transfer of substantial quantities of financial and technological assistance as a supplement to domestic effort of the developing countries, and such timely assistance as may be required."

The transference of financial and technological aid, allowing for the human failing that it is almost impossible to be completely altruistic, cannot be successfully monitored by national technology assessments. Once again, this is a U.N. principle which for effective application calls for the surveillance of an international OTA. If developed countries see a market for the exportation of some of their technological benefits and nearly all the concomitant pollution, the recipients in underdeveloped countries should be accurately advised on the real magnitude of the disbenefits they might accept. It is one thing to transfer a technological process to a country where the economic need is such that the social costs are, relative to the prosperous country, diminished. But even when the recipients are willing to accept greater social costs, they should really be certain about what they comprise. Caveat emptor is not a fair maxim to apply when the recipient has not sufficient expertise in technology and its assessment to know the dangers. Within the spirit of U.N. principle 1, the recipient must be protected.

Principle 10

"For the developing countries, stability of prices and adequate earnings for primary commodities and raw materials are essential to environmental management, since economic factors as well as ecological processes must be taken into account."

These economic factors will include the external social costs of the extraction and processing of the commodities and materials within the underdeveloped countries. Before agreeing to stabilize prices, those concerned would benefit from advice from an international OTA.

Principle 11

"The environmental policies of all States should enhance and not adversely affect the present or future development potential of developing countries, nor should they hamper the attainment of better living conditions for all, and appropriate steps should be taken by States and international organizations, to reach agreement on meeting the possible economic consequences of environmental measures."

Like national parliaments provided with technology assessments from only one source (the executive agencies of government), the U.N. will find it difficult usefully to guide agreement on averting or absorbing economic consequences of environmental measures without a counterbalancing technology assessment made independently by its own staff. (See Chapters 1 and 2 of this book.)

Principle 12

"Resources should be made available to preserve and improve the environment, taking into account the circumstances and particular requirements of the developing countries and any costs which may emanate from their incorporating environmental safeguards into their development planning and the need for making available to them, upon their request, additional technical and financial assistance for this purpose."

Within this principle, technical assistance to underdeveloped countries is best supplied through an international Office of Technology Assessment. Taxpayers in developed countries will more readily accept the allocation of resources for preservation and protection of the environment if they are presented with plausible

344

technology assessments (made without vested interest) and if they have a right of redressing any such expenditure through the provisions of legislation like NEPA. (See the previous chapters.)

Principles 13 and 14

"In order to achieve a more rational management of resources and thus to improve the environment, States should adopt an integrated and co-ordinated approach to their development planning so as to ensure that development is compatible with the need to protect and improve the human environment for the benefit of their population."

"Rational planning constitutes an essential tool for reconciling any conflict between the needs of development and the need to protect and improve the environment."

Technological assessment, within the definitions given in the preface, is obviously the tool to be used for assessments made *prior* to the rational planning approach recommended; the 102-impact statements of NEPA (see Chapter 2) are prime examples of assessments which could help to ensure protection of the environment. Naturally, the selection of a preferred plan depends upon the qualitative *and* quantitative definition of the development's benefit and the environmental and other costs. Yet it is not easy to quantify and agree upon an acceptable arrangement of all the relevant factors which, ideally, should be fed into a benefit—cost analysis (see Chapters 5 and 6). There is a glaring discrepancy between current competence in assessment and the U.S.'s aspirations for rational planning. Should the present swing away from an appreciation of technology continue and carry with it an associated reluctance to achieve numerate competence, this discrepancy (between plausible aspirations for rational planning and competence) will widen further. Rhetoric needs to be accompanied by willingness to acquire an understanding of the potential and limitations of assessment techniques (see Chapters 5—9). My present guess is that there are at least one hundred rhetoricians for every struggling analyst willing to pit himself against the redoubtable problems of quantifying the factors named by rhetoricians of the environment. With such disproportionate verbal incentive, unsupported by critical analytic competence, we should not be surprised if we are fobbed-off with some numerate quackery which could lead to the acceptance of harmful plans.

Principle 15

"Planning must be applied to human settlements and urbanisation to avoid adverse effects on the environment and to obtain maximum social, economic, and environmental benefits for all. In this respect projects which are designed for colonialist and racist domination must be abandoned."

As pointed out in Chapters 5 and 7, the maximization of social benefits is hampered by the difficulty of quantifying them. No generally acceptable method, or calculus, exists for handling social or environmental benefits and some disbelievers doubt the veracity of even economic analysis. Thus, once again, the rhetoric is perfectly plausible. But efficient means are awaited for first defining the social benefits, which are more usually inferred rather than quantified, so that maximal conditions may be found. I would, for example, like to see a quantifiable objective for the economic, social, and environmental disbenefits and benefits of an urban motorway.

Principle 16

"Demographic policies, which are without prejudice to basic human rights and which are deemed appropriate by Governments concerned, should be applied in those regions where the rate of population growth or excessive population concentrations are likely to have adverse effects on the environment or development, or where low population may prevent improvement of the human environment and impede development."

To obey this principle, the "Governments concerned" must have assessed the need for their demographic policies with more than normal care because the tyranny, which presages the most unbearable social disbenefits, can begin with forced movement of populations. Furthermore, the government must make the assessment of the demographic policy available to the affected populations and these populations should also have access to alternative assessments made outside the executive branches of government by an OTA (see Chapter 2) or a consultancy representing their interest (see Chapters 10—12).

Principle 17

"Appropriate national institutions must be entrusted with the task of planning, managing, or controlling the environmental resources of States to enhance environmental quality."

The brunt of such planning will, of necessity, be carried by the executive arm of government on their consultants. But independent assessment of these plans should also be carried out by skilled multi-disciplinary teams responsible only to elected parliaments and working in the public eye (see Chapter 2).

Management and control of environmental resources requires the active co-operation of the legal fraternity (see Chapter 2) and their assistance must be sought in drafting, enacting, and enforcing appropriate NEPA-type legislation. This will not be easy because lawyers are not particularly prone to legislate for actions which have not been experienced but may occur in the future. And the legal profession needs to be more heavily buttressed with technical expertise than it is at present; see Chapter 11 for apposite remarks by the British Lord Chancellor.

My specific recommendation is that the E.E.C. countries begin to plan for European versions of OTA and NEPA. In the face of the American evidence for the usefulness of such an Office and Act, the OECD (1971) study on "Science Growth and Society", and the need to plan European institutions for the future, whilst we are in a state of flux, such plans should not be delayed much longer.

Principle 18

"Science and technology, as part of their contribution to economic and social development, must be applied to the identification, avoidance, and control of environmental risks and the solution of environmental problems and for the common good of mankind."

At the risk of repeating myself, ad nauseam, it must be said that principle 18 enshrines the basic premise on which the concept of technology assessment is based. How good we become at technology assessment depends largely upon the firmness of our resolve and the depth of our pockets when we approach the subject. Experience suggests that quality assessment is the easiest thing to fake, but the hardest thing to produce, in benefit—cost analysis. In a post-industrial society, the assessment of technology could well cost more than its production. Since we seem bound to live with technology, let us determine to choose which kind.

347

Principle 19

> "Education in environmental matters, for the younger generation as well as adults, giving due consideration to the underprivileged, is essential to broaden the basis for enlightened opinion and responsible conduct.... in protecting and improving the environment...."

This point has been emphasized in Chapter 11 and though the U.S. Committee on Government Operations categorized the protagonists of principle 19 as "idealists", education is the best permanent defence against a ravaged environment. Chapter 10 describes some of the university precursors to more generalized education in environmental matters. Realists will appreciate that such education needs to contain instruction on how to do the associated "sums" and this means that elementary assessment techniques should be cohesively taught.

Principle 20

> "Scientific research and development in the context of environmental problems, both national and multi-national, must be promoted in all countries, especially the developing countries. In this connection, the free flow of up-to-date scientific information and experience must be supported and assisted, to facilitate the solution of environmental problems. Environmental technologies should be made available to developing countries on terms which would encourage their wide dissemination without constituting an economic burden on the developing countries."

It is rumoured that the U.N. will, after carefully digesting the fruits of the Stockholm conference, set up a computerized data bank of sources of environmental information. This move will partially satisfy principle 20 and exchanges of environmental information between countries via the data bank will be possible. But as an historical note, we should remember that the environmental impact statements made necessary by America's NEPA are already available to all at modest cost (see Chapters 2 and 12). Moreover, it should be recognized that scientific research, particularly curiosity-motivated research (COR), is extremely difficult to assess and it is from COR that discoveries arise which eventually become technological innovations.

Norton-Taylor (1972) and European Communities Bulletin (1971) indicate that the European Commission has proposed a radical new approach towards co-operation in research and development within the Common Market. If the suggestions of Signor Spinelli (the Commissioner responsible for industrial and technological affairs) that a European Research and Development Committee and a European Science Foundation devoted to COR be instituted are accepted, then the *European Communities have an early opportunity to establish an Office of Technology Assessment, an environmental policy act, and also help underdeveloped countries with environmental problems.*

The Communities could, with one re-organization, satisfy the U.N.'s principle 20, establish an OTA responsible to the European Parliament (but in communication with Community governments carrying out assessments within executive agencies), and give environmental protection to members of the community with a Europeanized version of NEPA (suitably watered down to protect national sovereignty). At a stroke, tardy governments could be forced to adjust Edwardian policies of environmental paternalism because the best protection against the imagined insults of assessment by external bodies is self assessment and National OTA's would soon emerge. What a pity it is that Lord Rothschild's "think-tank" (see Chapter 3) has not been asked to look at such an optimistically postulated idea for protecting, and creating, collective benefit!

Principle 21

"States have, in accordance with the charter of the United Nations and the principles of international law, the sovereign right to exploit their own resources pursuant to their own environmental policies, and the responsibility to ensure that activities within their jurisdiction or control do not cause damage to the environment of other States or areas beyond the limits of national jurisdiction."

Only an international OTA can provide the evidence necessary to support objectively an international judgement in a case of pollution beyond the polluter's national boundary. Appellants and defendants will naturally proffer their own technology assessments in such a case but, as with a State's own legal proceedings, the bench needs independent technical support. Recognition of sovereignty in environmental matters is no real excuse for a State refusing to give its citizens the benefits of environmental legislation pioneered elsewhere. I am very pleased indeed to note that the Rt. Hon. Peter Walker is on record as welcoming "working with our friends throughout the world to see if there is a total international approach to these problems" (see Chapter 2). As Secretary of State for the Environment he has, I hope, paid sufficient attention to the possibility of seeing whether the NEPA concept can be internationally transferred to the United Kingdom. One can hardly expect a member of government to call for the establishment of an OTA because it means there is some surveillance over executive agencies working to their instructions. For action on such matters within the United Kingdom I look towards the Parliamentary Select Committee for Science and Technology (see Chapter 3 for encouraging portends).

Principle 22

"States shall co-operate to develop further the international law regarding liability and compensation for the victims of pollution and other environmental damage caused by activities within the jurisdiction or control of such States to areas beyond their jurisdiction."

One possible form of co-operation has been mentioned under principle 21 of this chapter. A useful start would be to establish an international OTA.

Principle 23

"Without prejudice to such general principles as may be agreed upon by the international community, or to the criteria and minimum levels which will have to be determined nationally, it will be essential in all cases to consider the systems of values prevailing in each country, and the extent of the applicability of standards which are valid for the most advanced countries but which may be inappropriate and of unwarranted social cost for the developing countries."

It will not be easy to agree which standards used by the most advanced countries are best (see the note on the discrepancy between Russian and American standards for acceptable levels of exposure to microwave radiation in Chapter 10). The question of social cost of technology has been commented upon under principle 9; the developing countries may choose to accept certain social costs but, without the assistance of technology assessments, how can they be sure they have perceived them all?

350

Principle 24

"International matters concerning the protection and improvement of the environment should be handled in a co-operative spirit by all countries, big or small, on an equal footing. Co-operations through multilateral or bilateral arrangements or other appropriate means is essential to prevent, eliminate or reduce and effectively control adverse environmental effects resulting from activities conducted in all spheres, in such a way that due account is taken of the sovereignty and interests of all States."

A multilateral agreement is needed to set up an international Office of technology assessment. Logically the first task of such an OTA would be to define areas of the most propitious kind for co-operation between States.

Principle 25

"States shall ensure that international organizations play a co-ordinated, efficient, and dynamic role for the protection of the environment."

How? Through liaison co-ordinated by an international OTA.

Having tied (or hidden) the loose ends of this book under the twenty five principles, it may seem that I have placed undue emphasis in this end Chapter upon the concerns of the environmentalists. Industrialists should take note, however, that the definition of technology assessment used throughout this text contains the words "feasible technological alternatives and constructive opportunities". Industrial entrepreneurs wishing to maintain most of their existing profit flow and grasp ethical opportunities based on the new consciousness of need to tackle environmental problems, should have no unsurmountable difficulties in achieving success in a manner commensurate with collective social benefit. All they need do is resolve to work with, and not against, the flow of environmentalists' constructive action. Effort expended now in becoming familiar with present thinking on environmental problems and solutions based upon technology assessment will be well invested.

References

European Communities Bulletin (1971 No. 1). "Commision Memorandum to

351

the Council concerning Overall Community Action on Scientific and Technological Research and Development", *European Communities Bulletin No. 1*, 11 Nov., Brussels.

Landsberg, H.H. *et al.* (1960). *Resources in America's Future, 1960—2000*, New York, John Hopkins.

Medford, R.D. (1969). "Marine Mining in Britain", page 475, *Mining Magazine*, Vol. 121, No. 6, December 1969.

Norton-Taylor, R. (1972). "Radical Plan for EEC Research", *The Guardian*, June 15th, 1972, London.

OECD (1971). "Science Growth and Society. A New Perspective", *Report of the Secretary-General's Ad Hoc Group on New Concepts of Science Policy*, Paris: OECD.

INDEX*

Abramowitz, M. and Stegun, I.A.,
164, 174, 177r
Adams, John Gordon Underwood,
320, 327r
Advocacy, between analysts, 9
—, between scientists, 13
AD-X2, 36
Anti-ballistic missile (AMB), 27
Apportionment of benefit, 255,
256—265
Aquiculturalists, 208
Arbiters of policy, 1—8
—, relationships with analysts,
125—127
—, frustration of, 126
Arnfield, Ron, 4, 14r
Ash, Maurice, 324
Ashby, Eric, 21
Atomic Energy Commission, 21
Atomic reactor wastes, 21
Attitudes of T.A. staff, 116—118
Ayres, Robert, 208, 244, 245r

Baez and El Sum, 274
Baker, William O., 52 (quot.)
Barfield, C.E., 21, 22, 31, 56r
Battelle Memorial Institute, 2
Bauer, R.A., 235
—, on social indicators, 245r
Bell Telephone Laboratories, 275
Bellman, R., 162, 177r
Benefit-cost, 29, 124—177

Benn, Anthony Wedgewood, 99,
101 (quot.), 102r
Bill HR 10243 to establish an OTA
for U.S. Congress, 16
—, comments by European compa-
ny lawyer, 320
—, Edward Kennedy's evidence, 27
—, full text, 77—91, 113, 116
Bill S 2302 to establish an OTA for
U.S. Congress, 16
—, full text, 58—76
Biological effects of ELF magnetic
fields, 25
Biological growth model, 201
Boltzman's Law, 255
Bonse, Ulrich, 275
Bowers, Raymond, 316
Bray, Jeremy, 55
Bridgewater, A.V., 243, 245r, 296
British Alkaline Inspectorate, 55
Bronwell, Arthur, 314 (quot.), 317r
Brookes, Harvey, 39
Brookes, Jack, 41, 42 (quot.), 113
Brown, Murray, 256
Byatt, I.R.C., 212, 286r
Byatt-Cohen Model, 250—252

Cabinet Office, 99
Cade, Joe, 305, 317r
Calvert Cliffs, see Court decisions
Car exhaust pollution, 333
Carpenter, Richard, V, 247

* r = reference, quot. = quotation.

353

Catalytic men, 284
Central assessment agencies, means of canalising activists, 10
—, recommendations in OECD's publication "Science Growth and Society", 10
Central Policy Review Staff (CPRS), 6, 96, 104
CERN, 37
Cetron, Marvin, 198, 240, 245
le Chatelier, 49, 114
Chaumier, J., 5, 14r, 295, 303r
Cherwell, Lord (Professor Lindemann), 244, 296
Chief Scientist's responsibility in customer-contractor role play, 6, 98
Chorus, Claudius, 11, 14r
Civil assessment of technology, 2
Civil Service Staff College, 92
CLEAN, 28
Club of Rome, 10
Coal research, 37
Cobb-Douglas function, 156, 160
Cohen, A.V., 212, 286r
Coleman, J.S., 202, 245r
Computer Aided Instruction (CAI), 152, 155
Confederation of British Industries (CBI), 322
Congressional Research Service, 36
Consensus syndrome, 232
Contracts for Industry in TA, 330
Controllers of R & D, responsibility in customer-contractor role play, 6, 98
Corps of Engineers (U.S.), 25, 27
—, frustrated by NEPA, 28, 124
Corrigan, R., 31, 56r
Cost—effectiveness analysis, 96, 178—196
Council on Environmental Quality (CEQ), 17
Council for Scientific Policy, 6
Court decisions, under NEPA, 18
—, Calvert Cliffs Coordinating Committee vs. AEC, 21
—, Greene County Planning Board vs. Federal Power Commission, 21
—, National Resources Defense Council vs. Morton, 20
—, Wilderness Society vs. Morton, 31
Crampon, R.C., 17, 21, 50, 56r
Cross impact analysis, 240

Curiosity Motivated Research (COR), 247—287
Curlin, James, 48 (quot.)
Customer-contractor role playing, 6

Daddario, Emilio Q., 34, 35 (quot.)
Dainton, Fredrick, contribution to Green Paper, 6
Dalby, James F., 240
Dalkey, N., 238, 245r
Davis, John W., 41, 42 (quot.)
Defence organizations, 1, 96
Defoliant herbicides, 27
Degeneration syndrome, 295
Delayed-growth models, 205—206
Delphi technique,
—, description, 238
—, "two-legged", 115
—, usefulness, 115
Department of Defense (U.S.), 25, 37, 180
Department of Education and Science (U.K.), 250
Department of Liberal Studies in Science, Manchester University, 251, 311
Department of Trade and Industry, 108
Depletion syndrome, 340
Digital computers,
—, change of organization's utility with use of, 193
—, cost-effectiveness of, 180—194
—, and growth of TA, 5
Discrete probabilities, 134—141
Discounted cash-flow, 127—134
—, discontinuous vs. continuous cash-flow, 128
—, numerical method, 170
Disraeli, 54, 55
Dixon, Bernard, 325
Doomsday predictions, 11
Dupuit, J., pioneer of benefit-cost, 124, 177r

Economic impact of technology, 92, 95
Economist, 235
Electric vehicles, 39
Electron microscope, 253
Ellsberg, Daniel, 126
Energy crisis, 23
Environmental Defense Fund, 28
Environmental Impact Statements, NEPA, concept of, 17

—, where to buy, 23
Environmentalists, 26, 322—325
—, suggestions for programme, 325
Enzymes, 334
de l'Estoile, description and critique of his relevance technique, 211—227
Ethical quotient, 179, 236, 237
European Communities, 348
European Industrial Marketing Research Association, EVAF, 210, 298
European reaction to NEPA, 23
European versions of NEPA and OTA Acts, 347, 349
Exploratory technological forecasting, 3, 4
Exponential growth model, 203

Fisher, J.C., 202, 245r
Fluoridation, 38
Fontela, 228, 245r
Forrester, Jay, computer models by, 11, 14r, 307
Freedom of Information Act (U.S.), 17
Fulton Committee, influence, 3

Gabor, Dennis, 92 (quot.)
—, on ethical quotients, 236—237, 271, 273
Galbraith, J.K., 54, 56r, 97, 102r
Garmatz, A., 16 (quot.), 56r
Goals,
—, in French relevance system, 214
—, in national objectives, 228—231
—, mission-orientated, 4
—, social, 4
Goldman, David, teacher of T.A., 304, 311
Goldman, M.I., similarities between capitalist and communist treatment of pollution, 319, 328r
Gompertz function, 205
Goodman, C.S., 209, 245r
Governments agencies, propensity to withhold information, 9
Green, Harold, 46, 47, 56r
Green, Leon, 52
Green Paper, "A Framework for Government R & D", 6, 97, 101
Greenberg, Dan, 32
Gross National Product, 154, 155, 160
Gruber and Marquis, 249, 286r

Hamilton, P., 292, 303
Hart, Michael, 275
Heath, Edward, statement on opportunities for scientists, 311
Helmer, O., 238, 245r
Herbicides, 18
High-energy physics, 37
Hill, K.M., 113, 121, 305, 317r
Hill Samuel & Co., 104
Hobbs, John, 212
Holography, 265—283
Huddle, Franklin P., 24, 33, 34
—, magnum opus, 36, 56r, 233, 246r
Hudson Institute, 2

IBM, patent on X-ray microscope, 268, 271, 275
—, sponshorship of Program on Technology and Society, 312
Iconic models, 153
Industrial Intelligence, 140, 329—337
Industry, American and T.A., 51—54
Innovation (Journal), 52, 56r
Insecticides, 38
Intelligence activity,
—, EDP in, 204
—, ethical industrial intelligence, 288—303
—, interviews, 290
—, sod's law, 292
—, and technological forecasting, 4
Interactions between analysts, 112
Internal rate of return, 131
—, example, 132
International Institute of Applied Systems Analysis, 12
International Institute for Management of Technology, comments by Jantsch, 305
International OTAs, need for, 342, 343, 344
International Society for Technology Assessment (ISTA), 11
Invisible colleges, 8, 10—12

Jaeger, J.C., 129, 164, 177r
Jantsch, Erich, 2, 14r, 197, 228, 296, 305, 330, 337r
Jevons, Frederick, 251
Jolly Green Giant (with double hernia), 45
Jones, A., 110, 121r

Jones, M.V., 332, 337r
Jordan, Everett B., 41
de Jouvenal, 3, 14r

Kahn, Herman, 3, 14r, 237, 246r
Kantrowitz, Arthur, suggestions for
  increasing effectiveness of OTAs,
  13, 14r, 48
Karman, Von, 244
Katz, Milton, 45 (quot.), 150, 177r
Keeping up with the Jones (model),
  205
Kendall, M.G. and Stewart, A., 144,
  175, 177r
Kennedy, Edward M., on the ABM
  dispute, 27 (quot.), 315
Kiefer, David, 35, 53, 56r
Kirchner, E., 329, 337r
Knezo, Genevieve, 32, 35, 36, 45,
  50, 56r

Langrish, John, 248, 251, 286r,
  291, 308, 318r
Lansberg, H.H., 340
Laplace Transform, 129, 164
Laserson, N.A., 329, 337r
Lasky's evidence against the "depletion syndrome", 340
Law,
—, and technology assessment, 44—
  51
—, courts as sentinel of technology,
  49
Lenz, Ralph, 206, 246r
Lewandowski, Rudolf, 201, 204,
  246r
Limits to growth, see Forrester, 8,
  10
Lindberg, Charles A., 1 (quot.)
—, influence on OTA concept, 35
Logistic curves, see S-curves, 105
—, in decline of unethical intelligencer, 292—293

Management, good vs. bad, 136—
  140
Management Science, 2
Mansfield, E., 105, 121r, 137, 177r
Market research, in technology assessment, 114
Mathematical modelling,
—, in advising the President, 163
—, in benefit-cost analysis, 161—
  163
Maxwell Stamp Associates Ltd.,
  104

Mclaren, Peter, 251
Meadows, D.L., 8, 11, 14r, 160,
  307
Medford, R.D., 38, 56r, 113, 121r,
  143, 145, 147, 198, 206, 241,
  243, 246r, 261, 340, 352r
Methodology gap, 332
Microwaves, see Bowers, 316
Military projects, 27
Miller, 49, 57
Ministry of Technology, 5, 6, 7,
  104, 180, 265
Mitre Corporation, studies for OST
  on TA, 332
Mohole, 37
Morphological analysis, 240—244
Mueller, Robert Kirk, 12

Nader, Ralph, 9
NAE Committee on Public Engin-
  eering, 39
NAS Committee on Science and
  Public Policy, 39
NASA Programme, 232
National Committee for Education-
  al Technology (NCET), 152, 155
National Environmental Policy Act
  (NEPA), 15, 16—32, 314, 321,
  341, 345, 347, 348, 350
National Goals Research Staff
  (NGRS), 233
National Wildlife Federation, 22
New York Times, 126
Nixon, President, on NGRS, 233
  (quot.)
Normative Technological Forecast-
  ing, 3
Norton-Taylor, R., 348, 352r
N.R.D.C., and civil research, 5

Obscurantists of Theory, 145
OECD, European observer, 2, see
  Central assessment agencies, 56r,
  297
Office of Science and Technology
  (President's) (OST), 332
Office of Technology Assessment
  (OTA), American concept, 10,
  12, 32—44, 98
Offshore oil and gas leases, affected
  by NEPA, 20
Ogres in the scientific garden, 99
Operations Research, 2
—, American Society of, 27
Over-runs, in cost and time, 137

356

Pareto optimization, 180
Parker, Frank, Delphi survey of chemical industry, 239, 246r
Parliamentary Select Committee on Science and Technology,
—, debate on Lord Rothschild's proposals, 99
—, recommendations which auger well for U.K. formation of an OTA, 100
Parvin, R.H., 228, 246r
PATTERN, 214, 218
—, description of relevance technique, 227—232
Pay-back criteria, 132
—, example, 133
Pearce, Frank, 288, 289, 303r
Pearl, R., 201, 246
Pearson, Alan, 134
Pedlar, Kit (T.V. programme), 3
"Pentagon Papers", 126
Peter, L.J., 331, 337r
Platt, Brigadier General, 299, 303r
Politicians, procrastination in legislating for an OTA, 10
Pollack, Jackson, 153
Popov, Mikolai, 319 (quot.)
PPBS, 120
Present value, 127—134
—, probability of achieving, 134—141
Prest, A.R., 124, 177r
Programme on Technology and Society, Harvard University, 312
Programmes Analysis Unit (PAU), 6, 7, 49, 93, 103—123, 153, 180, 227, 241, 300, 331
Project Camelot, 37
Project Sanguine, Draft Environmental Impact Statement, 25
Proxy measures, 209
Pry, R.H., 202, 245r
Public involvement with TA, 319—328

Ralph, C.A., 198, 240, 245r
RAND Institute, 2, 126
Reed, Laurance, 107
Relevance techniques, 120, 211—237
Representative Esch, 43 (quot.)
Representative Fugua, 43 (quot.)
Representative Hanna, 44 (quot.)
Research and Development,
—, contractor, 5
—, evaluation of, 7

—, propensity to use computers, 181
Research Advisory Committees, 5
Roberts, Edward B., 151, 211
Rose, Hilary, 208, 246r
Roskill Commission, 320, 328r
Rothman, Harry, 311, 320, 328r
Rothschild, Lord, 6, 7, 93, 97, 100, 104
Royal Commission on Environmental Pollution, 24
Royal Society, 12
Ruckelshaus, William D., 31

Salam, Abdus, 247 (quot.)
Salk vaccine, 38
SAPPHO, 249, 286r
Scenario writing, 237
Schlesinger, J.R., 8, 14
Schlesinger, James, Chairman of the U.S.A.E.C., 22 (quot.), 32
Schon, Donald A., 245 (quot.)
Science Journal, 52, 56r
Scientific judges, 13
"Scientific method", 8
Scientists, dichotomy with politicians, 32
S-curves, 4, 201—206
Secretary of State and Industry, answer to L. Reed, 107
Section 102 of NEPA,
—, draft statements, 17
—, requirements leading to technology assessments, 16
—, requirements listed, 17
—, statements filed with CEQ, 18—20
Sectorial projects, 110
Sensitivity analysis, 151
Sierra Club, 22
Sigford, J.V., 228, 246r
Singular projects, 109
Smidt, S., model of utility of digital computer, 179, 196r
Snow, C.P., Lord, 32
Social benefits, example, 152
Social costs, internalization of, 32
Social goals, see Goals, 103
Social impact of technology, 92, 95
Social sciences, 37
Social time preference rate, 128
Société d'Economie et de Mathématique Appliquées, 212
Solla Price, Derek, 284
Spitz, Edward A., 161, 177r
Stanford Research Institute, 2

Starr, Chauncey, 39, 312
Sterling, Claire, 12 (quot.)
Strasser, Gabor, 51, 332, 337r
Stroke, G.W., 276, 286r
Studiengruppe für Systemforsch-
ung, 313
Sub-Committee on Science Re-
search and Development, 34
Submarines, 25
Subpena powers, 41, 113
Supersonic Transport (SST),
—, case study by Huddle, 39
—, Concorde, 95, 315
—, degeneration of stratospheric
ozone, 95
Systems analysis, military, 178,
305—307

Tanenbaum, Morris, 53 (quot.)
Taylor, Keith, 143
Technical Information for Con-
gress, 33
Technological forecasting,
—, defined by Jantsch, 3
—, extrapolation techniques, 198—
210
—, first European conferences, 4
—, normative techniques, 210—237
—, in technology assessment, 197—
246
Technological hypochondria, 96
Technology transfer, 36
Test Ban Treaty, 37
Thalidomide and politicians, 38
Think-tank, activities, 2
—, Lord Rothschild's CPRS, 97,
104
Threat phenomenon, 4
Tinker, J., 16 (quot.), 23, 56r, 321
TRACES, 249
Train, Russell, CEQ Chairman, 31
Trans-Alaska Pipeline,
—, agencies asserting statutory
rights, 31
—, and NEPA, 29
—, victory for environmentalists, 31
Tribe, Laurence, 50 (quot.), 56r,
336
Tucker, A., 99, 100, 102r
Turvey, R., 124, 177r

U.K.A.E.A., 5, 7, 104, 109
U.K. Civil Service, 93, 98

U.K. Public Service, T.A. in, 92—
102
U.K. Research Councils, 100
Undersea mining, 145
Undersea technology, 26
United Nations Conference on the
Human Environment, 303,
338—351
Universities,
—, systems universities, 305
—, teaching of T.A., 310
—, and technology assessment,
304—318
—, trans-disciplinary, 307
University of Europe, 315
Utility,
—, central, 186
—, curve, 182
—, law of diminishing returns, 182
—, of sub-group, 183
U.S. Committee on Merchant Marine and Fisheries, 16

Vested interest, technology assessment to combat, 8
Vietnam, 38

Walford, F., 179
Walker, Peter, 24 (quot.)
—, admiration of Disraeli the en-
vironmentalist, 54, 57r
Water, excess demand for in Eng-
land, 198
Water Resources Board, 200
Wiesner, Jerome, influence on TA
concept, 35
Wildlife, 341
Wilford, N., 28 (quot.), 57r
Wills, G., 4, 14r
Wood, E.S., 208, 246r

X-ray microscope, 250, 265—283
X-ray physics,
—, cardinal events, 270
—, foreseeable breakthrough, 267
—, prognostications on, 268

Yeager, Philip B., 35
Young, D.R., 48, 57r

Zaleski, E., 320, 328
Zwicky, Fritz, 240, 246r